Ternary Phase Diagr
Materials Science

Ternary Phase Diagrams in Materials Science

D. R. F. West

DSc, CEng, FIM

Imperial College of Science, Technology and Medicine
London, UK

and

N. Saunders, PhD

Thermotech Ltd
Surrey Technology Centre, Surrey Research Park
Guildford, UK

Third Edition

CRC Press
Taylor & Francis Group
Boca Raton London New York

CRC Press is an imprint of the
Taylor & Francis Group, an **informa** business

First published 2002 by Maney Publishing Ltd
Third edition published in 2002
Second edition published in 1982 by
Chapman and Hall Ltd

Published 2017 by CRC Press
Taylor & Francis Group
6000 Broken Sound Parkway NW, Suite 300
Boca Raton, FL 33487-2742

© The Institute of Materials 2002
CRC Press is an imprint of Taylor & Francis Group, an Informa business

First issued in paperback 2019

No claim to original U.S. Government works

ISBN 13: 978-0-367-44708-3 (pbk)
ISBN 13: 978-1-902653-52-5 (hbk)

Visit the Taylor & Francis Web site at
http://www.taylorandfrancis.com

and the CRC Press Web site at
http://www.crcpress.com

British Library Cataloguing in Publication Data
Available on application

Library of Congress Cataloging in Publication Data
Available on application

Typeset in India by Emptek, Inc.

Contents

FOREWORD

Since the publication of the first edition of *Ternary Equilibrium Diagrams* in 1965 there have been many important advances in the understanding of phase equilibria and phase transformations in materials systems. However, fundamental principles stemming from the work of J. Willard Gibbs more than a century ago remain foundational, and the field of phase diagrams, viz diagrammatical representations of phase relationships as a function of chemical composition, temperature, and, where appropriate, pressure, continues to be essential for many scientists and technologists.

Traditionally, undergraduate course for metallurgists, ceramists, chemists and geologists have included the theme of phase diagrams and their applications for the understanding and /or controlling the microstructure of heterogeneous systems, covering for example, the morphology and proportions of types of crystalline phases encountered. With the evolution and rapid development of the subject of materials science and engineering during the past half century, the range of applications of phase diagrams has been further extended to materials such as semiconductors, superconductors, composites and polymers.

Building on the single component systems, consisting of the elements of the Periodic Table, or compounds such as oxides, or polymers, there is a vast number of materials consisting of combinations of these components to be considered in phase diagram terms, involving liquids and solids (i.e. condensed phases) and gases. In many cases binary systems can be used effectively to represent important materials; outstanding examples include the iron-carbon, titanium-aluminium and alumina-silica diagrams. Two-component systems can, in many cases, be dealt with by plots of temperature vs composition, ignoring the pressure variable; moving to the three-component situation requires a three-dimensional representation of temperature and composition. However, the effort involved in acquiring experience in understanding and applying ternary phase diagrams in three-dimensions is rewarded by the considerable advantage of achieving a fuller analysis of more complex materials. For example, the iron-chromium-nickel, nickel-aluminium-chromium and alumina-magnesia-silica diagrams open up respectively important aspects of certain alloy steels, nickel-based superalloys, and industrial ceramics and slags; also these ternary systems can be used by suitable procedures to display important features of quaternary systems.

As was the case for the earlier editions of this book, the essential objective of this new edition is to serve a wide range of readers, particularly undergraduates and postgraduates and scientists already established in their careers, who wish to acquire or extend their understanding of ternary phase diagrams. The change in title of this edition from that of the previous editions *Ternary Equilibrium Diagrams* to *Ternary Phase Diagrams in Materials Science* recognises the extension of the subject matter covered.

This extended coverage includes discussion of a wider range of materials, notably through an expansion of the substantial Chapter on Selected Case Studies of industrial systems presented in the second edition of the book. These previously presented examples have been retained; they deal with systems, in some cases complex in nature, of practical relevance in various fields of metallurgy and materials science, including extraction metallurgy and materials processing, industrial alloys, ceramics and semiconductors. Particular emphasis is given to the use of phase diagrams as a means of understanding phase changes that occur as a function of changes in temperature; for example, for metallic alloys the processes of solidification and heat treatment are of special importance, while for ceramic systems, changes during heating processes such as sintering require special attention. In the present Edition the main new features of the Case Studies chapter are sections dealing with some aluminium alloys and titanium alloys, and with an important example of superconductors; also an example is discussed of the effect of partial pressure of oxygen on phase equilibria. In earlier chapters of this new edition, an additional feature is a brief introduction to some aspects of phase diagrams in polymeric systems.

Another feature of the present edition is a new Chapter which reviews at an introductory level some thermodynamic principles underpinning the understanding of phase diagrams. This forms the basis for presenting an outline of the CALPHAD approach for predicting phase diagrams from thermodynamic data. CALPHAD, in its some 40 years of development, has come to be a vital tool in reducing the extensive experimental work that was previously necessary to establish complex phase diagrams. Furthermore CALPHAD has the capability of dealing with multi-component systems in a way that was not previously feasible through traditional phase diagram representations; also the CALPHAD procedures support strongly the modern approach to the design of materials with specific structural features which can be achieved by control of composition and processing. The Chapter on Case Studies refers to examples of the application of CALPHAD to industrial situations; mention is also made of the important aspect of the quantitative analysis of non-equilibrium conditions resulting from rapid solidification.

In recent years much internationally organised activity has focused systematically on the critical evaluation of the large amount of experimental data reported in the literature; the aim of this activity, utilising also the CALPHAD approach, has been to provide the materials science community with the versions of a huge number of systems which can be regarded as the most reliable. In the interval since the publication of the Second edition of the present book, new information has led to changes in some of the phase diagrams considered as examples; this is notably the case in the Case Studies chapter. In the new Edition, however, in general, the phase diagrams have not been updated, although notes and references are provided in some cases to indicate the changes; it is hoped that the earlier versions of the diagrams will continue to serve well to illustrate the principles of interpretation.

Throughout the book the authors have done their best to present phase diagram drawings that comply with the relevant 'constitutional' rules. However, some of the diagrams may still have minor errors of construction. If readers detect any such inconsistencies, it would be appreciated if they could inform the authors.

The Second Edition incorporated, as a learning aid, a series of problems together with outline answers. These ranged from hypothetical examples to illustrate fundamental aspects to problems related to the actual materials systems dealt with in the Case Studies chapter; they have been retained and extended in the present Edition.

D. R. F. West and N. Saunders
February 2002

ACKNOWLEDGEMENTS

The authors wish to express grateful thanks for expert advice and assistance from the following: Dr A. T. Dinsdale and Professor B. B. Argent (the calculation and provision of the isothermal sections of the Fe-O-Si system), Dr Judith Driscoll (superconductor systems), Professor H. M. Flower (titanium alloys); Professor J. P. Neumann (The interpretation of liquidus projections in relation to invariant reactions) and Professor A. H. Windle, FRS, (Polymeric systems). Acknowledgements are also made to Dr Ursula Kattner, National Institute for Standards (NIST), Gaithersburg, USA, for reading a draft of Chapter 10 and providing valuable further information on sources of experimental data. The authors wish to add that, notwithstanding the valuable advice provided, any errors in the book are entirely their responsibility.

One of the authors (DRFW) also wishes to acknowledge again the contributions of those referred to in the Prefaces of the first two Editions of the book, whose advice and assistance have been carried through to this Third Edition: Professor M. C. Flemings, Professor F. P. Glasser, the late Dr Alan Prince, FREng, and the late Dr D. Lloyd Thomas

Permission to reproduce diagrams from various sources is also acknowledged at relevant points in the book. The cover illustration is used with acknowledgements to Dr A. T. Dinsdale and the National Physical Laboratory, Teddington. Also permission previously given for the second Edition by the University of London for examination questions to be reproduced is again acknowledged here.

Grateful appreciation is also expressed to Mr Peter Danckwerts at Maney Publishing for his expert and substantial assistance in the production of the book and to Jo Jacomb for her excellent work on the cover.

D. R. F. West
N. Saunders

PREFACE TO THE FIRST EDITION

Information concerning phase equilibria, such as can be represented by equilibrium diagrams is important in a number of fields of scientific study, and especially in metallurgy, ceramics, and chemistry. Materials of interest in these fields range through single-component to multi-component systems, the latter often being very complex. While many industrially important systems can be represented adequately by binary equilibrium diagrams, ternary diagrams provide a basis for studying a wide range of the more complex systems, such as are encountered in certain industrial alloys, and in slags and ceramics.

A number of texts dealing with ternary systems are already available (see for example references 2-8). Some of these are very comprehensive and include many diagrams, and their use will take the student of the subject to an advanced and detailed level of understanding. The present monograph is intended primarily as an introductory text, which it is hoped will prove useful for undergraduate and postgraduate students of metallurgy and ceramics, in particular. Special attention is given to the requirement of the metallurgist and ceramist to use phase diagrams as a means of understanding phase changes that occur during heating and cooling, as in solidification and heat treatment of alloys. This aspect is emphasized in terms of the principle of solidification reactions in relation to solid state constitution particularly making use of liquidus and solidus projections.

Hypothetical systems are considered, commencing with simple examples and then proceeding to those of greater complexity. The sequence is designed to show principles and is illustrated, in some cases, by reference to actual systems. The general treatment, in itself complete, nevertheless assumes a previous knowledge of binary phase equilibria.

The author wishes to thank Dr D. Lloyd Thomas for reading and commenting on the manuscript.

D. R. F. West
1965

xi

PREFACE TO THE SECOND EDITION

The second edition of this monograph retains the aim of the original text to provide an introductory treatment of ternary equilibrium diagrams. However, two significant modifications have been made to assist the reader in obtaining a detailed understanding of the subject and to emphasize the practical importance of ternary diagrams. These modifications are the provision of problems with outline solutions and answers and the addition of a substantial chapter to extend the coverage of actual systems.

The actual ternary systems, presented as 'case studies' in this chapter, are of practical relevance in various fields of metallurgy and materials science such as extraction and process metallurgy, industrial alloys, ceramics and semiconductors. Some of these systems show considerably greater complexity than those considered in the previous chapters, and they include examples of solid state transformations. The general approach in the new chapter is to outline the main features of the systems and to relate them to practical applications. It is hoped that the sequence chosen for the presentation of the examples will be helpful in the study of complex systems.

The author wishes to acknowledge gratefully the valuable advice and assistance provided by Mr Alan Prince, Dr D. Lloyd Thomas, Professor M. C. Flemings and Dr F. P. Glasser. Acknowledgement is made to the University of London for permission to reproduce examination questions.

D. R. F. West
1981

1. General Considerations

1.1 INTRODUCTION

The thermodynamic work of Josiah Willard Gibbs, carried out in the USA in the latter part of the 19th century, provided the fundamental foundation for the understanding of phase equilibria and for the representation of such equilibria in the form of equilibrium (phase) diagrams. In 1876 Gibbs published the first half of his great memoir *On the Heterogeneous Equilibria of Substances* in the *Transactions of the Connecticut Academy of Arts and Sciences,*[1] and the second half followed two years later. This work developed and virtually completed the theory of chemical thermodynamics, and provided basic theory for the development of physical chemistry. Also, in metallurgy Gibbs' work provided an essential basis and guiding principle for the extensive interest in phase diagram determination and application that occurred as a major feature of the evolution of physical metallurgy around the end of the 19th century. As the 20th century progressed, experimental investigations of phase diagrams and phase transformations formed a major feature of metallurgical activity, related to alloy development and processing; in addition there has been extensive work on non-metallic systems, such as ceramics, while more recently polymeric systems have also become of considerable importance. In the latter half of the 20th century, the use of thermodynamic data and procedures for the calculation of phase diagrams (CALPHAD) has been established as a vital field.

The development of the field of phase diagrams has been accompanied by the publication of a number of important textbooks, which overall provide wide coverage of topics, including the interpretation of ternary and higher order systems.[2-14] The Phase Rule as propounded by Gibbs is an essential feature in relation to the representation and interpretation of phase diagrams. Reference can be made to other sources e.g. Refs. 4,9, for accounts of the derivation of the Phase Rule and the assumptions involved, and a brief outline of its significance is given below.

1.2 THE PHASE RULE

A particular state of equilibrium is characterized by the number and identity of the *phases* present, and the Phase Rule expresses the relationship, at equilibrium, of the number of phases, P, with the number of components, C, and the number of degrees of freedom, F. The rule may be expressed as follows:

$P + F = C + 2$

Table 1.1

P	1	2	3	4
F	3	2	1	0

The *phases* are the homogeneous parts of a system, bounded by surfaces and, in principle at least, can be mechanically separated from one another. The term phase is often used less precisely in practice to include portions of a system that are not entirely homogeneous, and are hence not strictly in equilibrium e.g. phases in solid alloys often show compositional variations.

The *number of components* is the smallest number of independently variable constituents necessary for the statement of the compositions of all the phases in the system; in alloy systems, for example, the number of constituent metals.

The *number of degrees of freedom* of the equilibrium state is the number of conditions that can be altered independently without changing the state of the system, or which have to be specified to define completely the state of the system. In the above statement of the Phase Rule it is assumed that temperature, pressure and composition are the only externally controllable variables that influence the phase equilibria. In ternary systems there are two compositional variables to consider, since a statement of the concentrations of two components is sufficient to define the composition, e.g. in an alloy of metals A, B, and C, the specification of the percentages of, say, A and B suffices to establish the alloy composition.

In many instances in metallic systems, the vapour pressures of the liquid and solid phases are negligible, or small, in comparison with atmospheric pressure; the pressure may then be considered as a constant, and the Phase Rule reduces to

$P + F = C + 1$

One degree of freedom is exercised by fixing the pressure in this way. Applying this 'reduced rule' to a ternary system ($C = 3$), the general position is summarised as shown in Table 1.1. The present account concentrates mainly on those systems in which the pressure variable can be neglected.

1.3 THE TERNARY SPACE MODEL

To represent completely the phase equilibria at constant pressure in a ternary system, a three dimensional model, commonly termed a space model, is required; the representation of composition requires two dimensions, and that of temperature, a third dimension. The model used is a triangular prism (Figure 1.1), in which the temperature is plotted on the vertical axis, and the composition is represented on the base of the prism, which may be conveniently taken as an equilateral triangle.* Thus, in Figure 1.1,

---

* Gibbs reported a method of displaying the composition of ternary systems in the form of an equiliateral triangle; this method was rediscovered by Sir George Gabriel Stokes and was used in research on ternary alloys published by C.R.A. Wright and C. Thompson in a series of papers in the *Proceedings of the Royal Society* beginning in 1899.

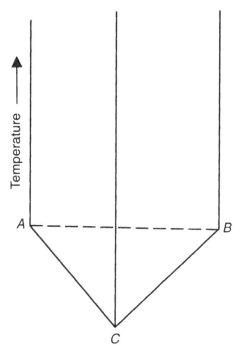

Fig. 1.1 Space model used for representing ternary systems.

the vertical sides of the prism represent the three binary systems *AB*, *BC*, and *AC* that make up the ternary system *ABC*.

Regarding the composition triangle the three corners represent the three components, points on the sides represent binary alloys, and points within the triangle ternary alloys. Generally, the length of each side of the triangle is considered as divided into 100 equal parts each of which represents 1% on the composition scale in each of the binary systems. Composition scales are commonly plotted in weight percentages. Then, the percentage by weight of each component in a ternary alloy (e.g. alloy *x* in Figure 1.2) is obtained by the following construction: a line *mn* is drawn parallel to side *BC* and the percentage of *A* is then given by the lengths *mB* or *nC*. The percentage of *B* is given by the lengths *oA* or *pC*, obtained by drawing line *op* parallel to side *AC*. Similarly, the percentage of *C* is obtained by drawing *rs* parallel to side *AB* and determining the lengths *rA* or *sB*.

Thus

%*A* = *mB* = *nC* = *xs* = *xp*
%*B* = *oA* = *pC* = *xr* = *xn*
%*C* = *rA* = *sB* = *xo* = *xm*

To facilitate the plotting of composition points, the equilateral triangle may be ruled with lines parallel to its sides (Figure 1.3).

Consider the plotting of the composition point of an alloy *y*, containing 50% *B* and 30% *C*. Starting from point *A*, a length *Ab* corresponding to 50% *B* is marked out along *AB*, and

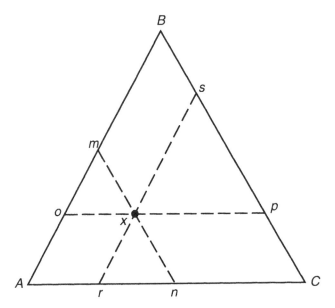

Fig. 1.2 Equilateral triangle to represent compositions.

then a length Ac corresponding to 30% C is marked out along AC. The composition point of the alloy then lies at the intersection of the appropriate ruled lines as shown.

An alternative, but less convenient method of representing the composition on the equilateral triangle is to take the altitude of the triangle as 100%, in which case the percentages of the individual components in a ternary alloy are given by the heights of the respective perpendiculars from the composition point to the opposite sides of the triangle.

A right-angled triangle is sometimes used to represent compositions. In the example shown in Figure 1.4 the percentage of A is obtained by subtracting the percentages of $B + C$ from 100. This method of representation is useful when interest is centred on alloys of metal A, containing relatively small amounts of B and C, that is when the representation of the A-rich corner of the system is required, as is often the care in aluminium-base alloys.

The following features are of interest in connection with compositional aspects of ternary systems:

1. If a straight line such as Am in Figure 1.5 is drawn from one corner of the triangle (i.e. from component A) to intersect the opposite side (i.e. BC), then the ratio of the other two components (i.e. B and C) is constant in alloys represented by points lying on the line. Considering alloy compositions along Am, as the composition point becomes progressively further away from A the percentage of A present in the alloys decreases, but the relative amounts of B and C remain the same.

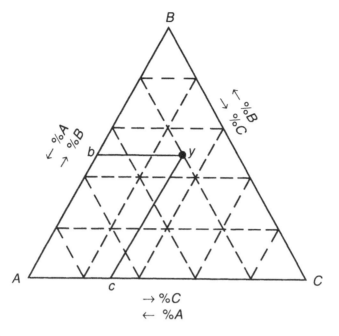

Fig. 1.3 Plotting alloy compositions on the compositional triangle.

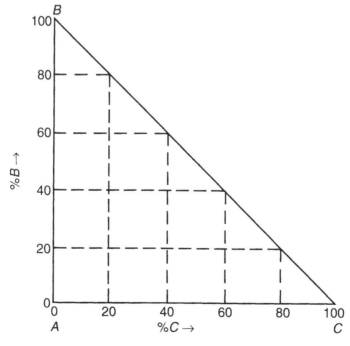

Fig. 1.4 Right angled-triangle to represent compositions.

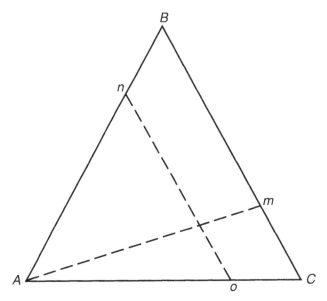

Fig. 1.5 Lines representing specific compositional characteristics.

2. A line drawn parallel to one side of the triangle is a line along which alloys contain a constant percentage of one component, i.e. along *no* (Figure 1.5) the percentage of *A* remains the same.

3. If two alloys (e.g. *d* and *e* in Figure 1.6) are mixed together, the composition *f* of the resulting mixture lies on a straight line joining the compositions of the two alloys. The total composition lies at the point dividing the line into segments inversely proportional to the ratio of the quantities of the original alloys. Then, using weight percentages:

$$\frac{Weight\ of\ d}{Weight\ of\ e} = \frac{ef}{df}$$

$$\%d = \frac{ef}{de} \times 100$$

$$\%e = \frac{df}{de} \times 100$$

This is the so-called 'lever principle' or 'lever rule' which finds an important application in 'tie-lines' when an alloy decomposes into a two-phase mixture (as discussed later).

4. If three alloys *x, y,* and *z* are mixed together, the composition of the mixture, *t* (Figure 1.6), lies within the triangle formed by joining the compositions of the alloys

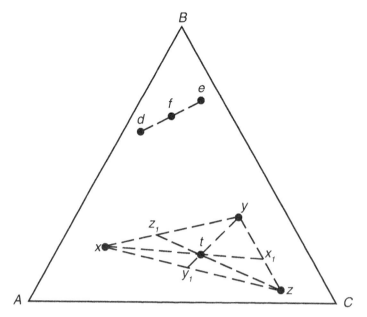

Fig. 1.6 Mixtures of two or three alloys or phases.

by straight lines. Using the 'centre of gravity' principle, the triangle may be considered as being supported on a point fulcrum at t and having masses at its corners proportional to the amounts of the alloys. Then, in equilibrium, the position of t is such that the percentages of the three alloys in the mixture are given by

$$\% \, x = \frac{x_1 t}{xx_1} \times 100$$

$$\% \, y = \frac{y_1 t}{yy_1} \times 100$$

$$\% \, z = \frac{z_1 t}{zz_1} \times 100$$

The application of this principle in 'tie-triangles', representing three phases in equilibrium, will be discussed later.

In general, except for relatively simple cases, it is difficult to depict clearly by means of perspective diagrams the complete ternary space model with its constituent spaces, curves, and points. It is convenient to use projected views (for example, a projection of the liquidus or the solidus surface on to the base of the space model), and also plane sections taken either horizontally (i.e. isothermally) or vertically through the space model. These types of diagrams are very useful when considering the solidification processes of alloys, or solid state transformations relevant to heat treatments, and are widely used for the presentation of experimental data for ternary systems.

1.4 PHASES AND THEIR EQUILIBRIA

This book deals primarily with equilibria between liquid and solid phases in ternary systems, mainly of the metallic and ceramic types. Chapter 2 and other chapters also refer to fundamental features of binary systems. In the liquid state, the components of many systems are completely soluble in one another in all proportions and therefore form a complete range of solutions. In cases where the components are not soluble in one another in all proportions there exists a region, or regions, in which two or three liquids co-exist; such a region can be termed a 'miscibility' or 'solubility gap'.

In the solid state, the phases are either the pure components of the system, solid solutions, or intermediate phases. In a substitutional solid solution of a metallic or ceramic system, each type of atom involved occupies positions on a common lattice, characteristic of the solvent element, while an interstitial solution has the solute atoms located in the interstices of the lattice of the solvent. In a ternary solution, one type of atom may be located substitutionally and another interstitially. Complete solid solubility is shown in some cases, but partial solid solubility is more common.

Many metallic and ceramic systems contain one or more intermediate phases (compounds) that occur over certain ranges of composition and temperature; such phases generally differ from the components with respect to lattice structure. Some intermediate phases are stable up to their melting point, while commonly others form from the liquid by a peritectic reaction; in alloy systems several types may be distinguished on the basis of atomic bonding and compositional ranges. In some ternary systems, the only intermediate phases present may be those derived from one or more of the binary systems, but there are many cases where true ternary phases occur, i.e. those whose composition ranges do not extend into the binary systems. When several intermediate phases occur, the phase diagrams may be complex because of the large number of possibilities for the equilibria between the phases. In particular, many ceramic systems show very great complexity in this respect.

Only a brief account is given of polymeric systems, although phase relationships are of considerable significance for these materials.(e.g. Refs. 17–19) The variables of composition, temperature and pressure are common to metallic, ceramic and polymeric systems. However, it is appropriate to note some of the relevant differences arising from the nature of polymers as consisting of long chain molecules, basically of carbon and hydrogen. In essence, polymer molecules are big, often with molecular weights in excess of 100,000. Considering polymer blends in comparison with metallic or ceramic alloys, there are important differences in terms of thermodynamic factors, which are discussed more fully in Chapter 2. A key point is that in polymers the entropy of mixing is 'per particle' i.e. per molecule, and the entropy of mixing term is thus much smaller than in the case of metals (there are simply fewer particles per unit volume); the result is that the heat of mixing tends to be the dominant term, and polymer blend systems are often immiscible over most of the composition range.

Polymers are complicated by other factors. In the solid, glassy state they are not in equilibrium, and where there is partial crystallinity the chains in the non-crystalline component are highly constrained by the crystals. Finally, as there is inevitably a distribution of molecular weights in the chains, the rigorous definition of a 'phase' according to Gibbs (see page 2) may be difficult to apply.

In Chapter 2, further comments on thermodynamic aspects are made, in relation to the phenomenon of phase separation giving rise to miscibility gaps in phase diagrams for solutions of polymers in various solvents, including other polymers. Reference is also made to liquid crystalline polymers in binary systems. Miscibility gaps in ternary polymeric systems are illustrated schematically in Chapter 4.

It is relevant to consider briefly, in the light of the Phase Rule, the representation of equilibria involving various numbers of phases. For this purpose it is assumed that the solid phases are of variable composition, and comparison is made of equilibria in binary and ternary systems, making reference to a hypothetical binary eutectic system showing partial solid solubility (Figure 1.7). It should be noted that constant pressure is assumed and the reduced Phase Rule is used.

1.4.1 ONE PHASE IS PRESENT

In the binary system, when only one phase is present there are two degrees of freedom, and to define the state of the system both the temperature and the concentration of one component must be specified; in the binary alloy diagram an area represents the range of existence of the phase (e.g. the liquid phase region in Figure 1.7). In a ternary system, application of the Phase Rule shows that there are three degrees of freedom; thus it is necessary to define the temperature and the concentrations of two components to define the state. The phase region is represented by a three-dimensional space.

1.4.2 TWO PHASES CO-EXIST

When two phases co-exist in equilibrium in the binary system, as for example in the L + β region of Figure 1.7, there is one degree of freedom. The two-phase region may be viewed as being made up of a stack of tie-lines (such as $L_1\beta_1$): each of these isothermal lines joins the composition points of the two phases in equilibrium at a particular temperature. The ends of the tie-lines at the various temperatures constitute the liquidus and solidus curves respectively. Thus, the composition points for each of the phases form a curve and the equilibrium is defined if one variable is stated, i.e. either the temperature, or the concentration of a component in one of the phases.

In a ternary system, when two phases co-exist, there are two degrees of freedom and two variables must be defined, e.g. the temperature, and a composition term for one of the phases.

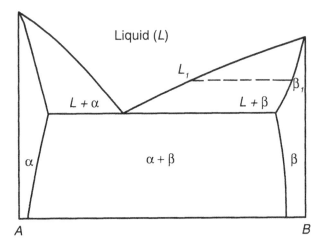

Fig. 1.7 Binary eutectic system showing partial solid solubility.

The two-phase region in the diagram is a space that may be considered as being filled with tie-lines whose extremities constitute the two surfaces bounding the space. Thus, the composition of each of the phases in equilibrium lies on one of these surfaces, not on a curve as in the binary system; this is a consequence of the extra degree of freedom in the ternary system as compared with the binary.

1.4.3 Three Phases Co-Exist

In the binary system, when three phases co-exist in equilibrium, there are no degrees of freedom, i.e. the system is invariant. Thus, at the eutectic temperature in Figure 1.7, the compositions of each of the phases, liquid, α and β, respectively, are represented by fixed points and the equilibrium is depicted in the diagram by an isothermal line.

In a ternary system, one degree of freedom remains when three phases co-exist, and the state is defined if either the temperature or a composition term for one of the phases is specified. The three-phase region in the ternary diagram is a space bounded by three surfaces and three curves formed by the intersection of pairs of these surfaces. An isothermal section through this space is a tie-triangle, as shown for example in Figure 1.8 for equilibrium between a liquid, L, and two solid solutions α and β. The sides of such a triangle are tie-lines, and the three corners represent the compositions of the three co-existing phases.

The three-phase space can be considered as being made up of a stack of such tie-triangles; the positions, at various temperatures, of the tie-lines bounding the triangle, mark out the boundary surfaces ('ruled' surfaces) of the region and the corresponding corner positions of the triangle mark out the boundary curves. Thus, the compositions of each of the three phases

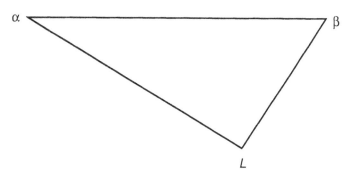

Fig. 1.8 Tie-triangle representing the equilibrium between liquid (L) and two solid solutions α and β.

are located along curves in the space diagram, as compared with their representation by points for the case of three-phase equilibrium in a binary system. Because of the additional degree of freedom in the case of a ternary system, three-phase reactions (e.g. of the eutectic or peritectic types) occur over a range of temperatures, and not at a fixed temperature as in a binary system.

1.4.4 FOUR PHASES CO-EXIST

With binary systems, the possible co-existence of four phases arises if pressure is taken into account as a variable, and if the complete diagram of temperature, pressure, and composition is considered.

In ternary systems, applying the reduced Phase Rule, the co-existence of four phases is a case of invariant equilibrium since there are no degrees of freedom. Such an equilibrium (e.g. of the ternary eutectic and peritectic types) exists only at a particular temperature in ternary systems, and the compositions of each of the four phases lie at fixed points on a plane in the space model corresponding to the particular temperature.

1.5 CLASSIFICATION AND FEATURES OF TERNARY METALLIC AND CERAMIC SYSTEMS

It is convenient to consider types of systems containing various numbers of phases, and on this basis the classification given below is adopted in the following pages; this classification also defines the types of phases and reactions dealt with in Chapters 3–7. Polymeric systems are referred to in Chapters 2 and 4.

- Chapter 3: Systems containing only two phases, namely a liquid + a solid solution.
- Chapter 4: Systems containing three phases (e.g. a liquid + 2 primary solid solutions) and showing a three-phase solidification reaction.
- Chapter 5: Systems containing four phases (e.g. a liquid + 3 primary solid solutions) and showing a four-phase invariant reaction.
- Chapter 6: Systems containing more than four phases (e.g. a liquid + 3 primary solid solutions + one or more intermediate phases) and showing two or more invariant reactions unless stated otherwise.

In general, complete miscibility in the liquid state is assumed; however, a liquid miscibility gap, with an associated monotectic-type reaction is briefly discussed in Chapter 4. Consideration of phase transformations in these systems is based on the assumption that equilibrium conditions are attained unless stated otherwise.

The approach adopted in this book is to focus initially on systems in which a significant degree of solid state solubility is an important feature, typifying the majority of metallic alloys. The case of negligible solid solubility, which is common in ceramic systems, is also introduced, beginning in Chapter 5. Specific examples of complex ceramic systems are included in Chapter 8, where also an outline is given of some differences in terminology which are found in the literature on ceramic systems. The transformations considered in metallic and ceramic systems are primarily those in which a liquid phase is concerned; typically the transformations are discussed in the context of solidification processes for metallic alloys and melting in ceramic systems. As far as the solid state regions of phase diagrams are concerned, these are not discussed in detail; the basic principles of interpretation for solidification can be applied, bearing in mind the slower diffusion rates as compared with liquids, but reference is made in Chapter 8 to allotropic transformations and eutectoid reactions in alloy steels.

An important point to bear in mind in the construction of phase diagrams is that the requirements of the Phase Rule lead to certain restrictions concerning the association of phase regions in space models and isothermal and vertical sections. This topic has been fully discussed by Prince,[9] including, for example, a generalised relationship concerning the dimensions of boundaries between adjoining phase regions.

1.6 PROBLEM

In a system *ABC*, a ternary alloy of composition 30 wt% *B* and 30 wt% *C* consists at a particular temperature of three phases of equilibrium compositions as follows:
- Liquid phase 50% *A*, 40% *B* and 10% *C*
- α solid solution 85% *A*, 10% *B* and 5% *C*
- γ solid solution 10% *A*, 20% *B* and 70% *C*
a. Calculate the proportions by weight of liquid, α and γ present in this alloy.
b. For the same temperature, deduce the composition of the alloy which will consist of equal proportions of α and γ phases of the compositions stated above, but with no liquid phase.

2. Some Thermodynamic Considerations for Phase Equilibria and Transformations

2.1 INTRODUCTION

The subject of thermodynamic stability in materials systems is of fundamental importance, underpinning considerations of phase equilibria and phase transformations. Considering a system at constant temperature and pressure, the relative stability is dependent on the Gibbs energy, G defined as:

$$G = H - TS \qquad (1)$$

where H is the enthalpy, T the absolute temperature, and S the entropy of the system. The enthalpy relates to the heat content of the system and is expressed as

$$H = E + PV \qquad (2)$$

where E is the internal energy of the system, P the pressure and V the volume.

E is associated with the total potential and kinetic energy of the atoms that constitute the system. During a transformation, the heat absorbed or evolved will depend on the change in E. The term PV takes account of volume changes and, if pressure is constant, the heat evolution or absorption is then directly equivalent to the change in H. In the context of the majority of situations considered in this book, dealing with solids and liquids (condensed phases), the PV term is generally very small compared to E, so that H is $-E$. The entropy, S, represents a measure of the randomness of the system.

The state of equilibrium, viz. that which is most stable, in a closed system having fixed mass and composition at constant pressure and temperature is that which has the lowest possible value of G.

Considering eq. (1), the state of equilibrium will depend on the relative values of H and S for the phases of interest. Solid phases are most stable at relatively low temperatures, since strong atomic binding leads to low internal energy and enthalpy. On the other hand, the stability of liquids and gases at high temperatures is due to the increased freedom of atomic movement, hence the dominance of the $-TS$ term.

Subsequent chapters of the present book emphasise the representation of ternary phase diagrams showing the equilibrium situations as a function of temperature and composition. However, the theme of phase transformations is also considered in some detail, particularly those involving solidification and melting processes, with special reference to illustrating changes in compositions and relative proportions of the phases as a function of temperature. Some discussion is also presented on

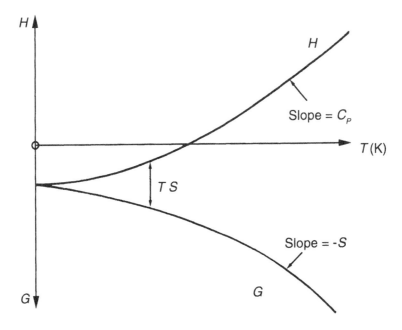

Fig. 2.1 Schematic variation of enthalpy and Gibbs energy with temperature for a pure metal in the solid state at constant pressure. (Reproduced by permission of Nelson Thornes Ltd., from D. A. Porter and K. E. Easterling, *Phase Transformations in Metals and Alloys*, second edition, 1992).

departures from stable equilibrium e.g. in relation to rapid solidification including the formation of metastable phases. Conditions of metastability correspond to a structural configuration representing a minimum condition of Gibbs energy. However, this configuration may not be the lowest possible value of G and there is an activation energy barrier to be overcome for transition to the stable state.

Considering a single component system, such as a pure metal at constant pressure, the variation of Gibbs energy with temperature in the solid state increasing from absolute zero, and its dependence on changes in H and S are illustrated qualitatively in Figure 2.1. The occurrence of a phase change (not shown in the Figure) such as melting of the solid, is associated with a discontinuity in the plot of G vs. temperature. When solid phases can consist in several different crystal structures, the Gibbs energy vs. temperature curves can be constructed for each of these structures. The intersection of the respective curves then corresponds to the equilibrium temperature for the occurrence of an allotropic transformation (Figure 2.2). The solid state allotropic transformations in Fe and Ti are utilised to their fullest extent in the control of microstructure of important industrial alloys based on these elements.

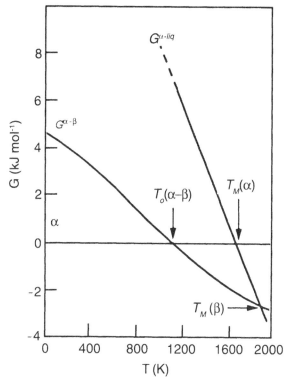

Fig. 2.2 Gibbs energy curves for α and β and liquid titanium. (Reprinted from L. Kaufman, *Acta Materialia*, 1959, **7**, 575, with permission from Elsevier Science).

2.2 BINARY AND TERNARY SOLUTIONS

In considering the effect of composition as a variable in metallic and ceramic systems, it is appropriate to first consider the case of solutions, both liquid and solid, formed in binary systems. Calculations can be made for the Gibbs energy of solutions based on various types of model. Assume that a homogeneous solid solution is formed between two components A and B, whose molar free energies are G_A and G_B respectively. There is a change in Gibbs energy, ΔG_{mix}, when the solution is formed:

$$\Delta G_{mix} = \Delta H_{mix} - T\Delta S_{mix} \tag{3}$$

where ΔH_{mix} is the heat of solution and ΔS_{mix} is the entropy difference between the mixed and unmixed conditions. For a solid solution consisting of atoms of types A and B, entropy is related to randomness (disorder) of the system by the Boltzmann equation

$$S = k \ln W \tag{4}$$

where k is Boltzmann's constant and W is a measure of randomness (disorder) of the system. In a solution, account must be taken of configurational entropy, representing the number of ways of arranging the atoms. Consider a crystal with N lattice sites, where N is Avogadro's Number, of which n are occupied by A atoms and $N\text{-}n$ by B atoms, then the number of ways of distributing the atoms is:

$$W = \left(\frac{N!}{n!(N!-n!)} \right) \tag{5}$$

hence

$$S = k.\ln\left(\frac{N!}{n!(N!-n!)} \right) \tag{6}$$

This leads mathematically, through use of Stirling's approximation, to the relationship

$$S = -Nk(x_A \ln x_A + x_B \ln x_B) = -R(x_A \ln x_A + x_B \ln x_B). \tag{7}$$

This defines the ideal entropy change on mixing. For the case of an ideal solution, where no repulsive or attractive interactions are considered to occur between the components, i.e. $\Delta H_{mix} = 0$, eq. (3) becomes

$$\Delta G_{mix}^{ideal} = RT(x_A \ln x_A + x_B \ln x_B) \tag{8}$$

The molar energy Gibbs energy per mole of an ideal solid solution is shown by the respective curves in Figure 2.3(a).

The concept of an ideal solution is, in reality, only an ideal and usually only useful for phases such as the gas phase. In condensed matter, there are invariably significant repulsive or attractive interactions between components. Therefore, an extension of the approach is needed and the simplest procedure for dealing with such interactions is by use of the regular solution model, where the excess heat of mixing ΔH_{mix}^{xs} is given by,

$$\Delta H_{mix}^{xs} = x_A x_B \Omega \tag{9}$$

where Ω is the regular solution interaction energy parameter and is related to the interaction energies of the bonds between the A and B components. A positive value of Ω corresponds to repulsive interactions while a negative value corresponds to attractive interactions. Hence equation (3) becomes

$$\Delta G_{mix} = x_A x_B \Omega + RT(x_A \ln x_A + x_B \ln x_B) \tag{10}$$

In practice Ω is also considered to contain an excess entropy term such that,

$$\Omega = A \text{ - } BT \tag{11}$$

where A and B are, respectively, temperature independent and dependent regular solution terms, in this case A being equivalent to ΔH_{mix}^{xs} while B is equivalent to ΔS_{mix}^{xs}. Equation (9) is then given in terms of an excess Gibbs energy of mixing

$$\Delta G_{mix}^{xs} = x_A x_B (A \text{ - } BT) \tag{12}$$

and eq.(10) becomes

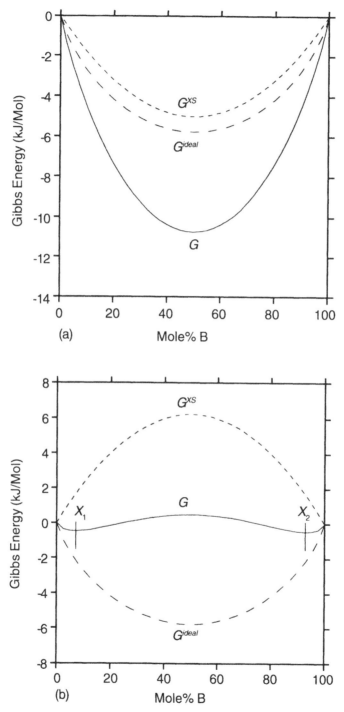

Fig. 2.3 Gibbs energy vs composition curves showing the effect of (a) attractive and (b) repulsive atomic interactions. (N. Saunders and A. P. Miodownik, *CALPHAD: Calculation of Phase Diagrams.* 1996, p.47; with permission from Elsevier Science).

$$\Delta G_{mix} = \Delta G_{mix}^{xs} + RT(x_A \ln x_A + x_B \ln x_B) \tag{13}$$

In the Fe–Ni system the *FCC* phase forms a solution with a negative value for Ω, meaning that Fe and Ni interactions are attractive. This, with the ideal entropy, leads to a smoothly changing curve at any temperature T, with a single minimum in the plot of Gibbs energy vs composition. (Figure 2.3a). In this case a continuous fcc solid solution is formed.

For a system such as Cu–Ag, Ω is strongly positive meaning that Cu and Ag interactions are repulsive. In this case, a Gibbs energy curve with two minima results, one at the Cu-rich end and the other at the Ag-rich end (Figure 2.3b). Alloys with compositions lying between x_1 and x_2 can then lower their energy by forming two phase structures, consisting of solid solutions of Ag in Cu and Cu in Ag respectively. Thus, the system shows a miscibility gap in the solid state. If a common tangent to the two minima in the Gibbs energy-composition curve is constructed, then its contact points, x_1 and x_2, with the Gibbs energy curve give the compositions of the two solid solutions at this particular temperature.

In the case of ternary solutions, the Gibbs energy vs. composition curves are three dimensional in form. This is because of the additional degree of freedom and means that the tangent construction applied when three phases are in equilibrium (tie-triangle instead of tie-line) involves a tangent plane common to the minima on the Gibbs energy curves.

A further illustration of a solid state miscibility gap, and the associated Gibbs energy curves is shown in Figure 2.4(a) and (b). The system shows, at high temperature, a single phase region of solid solution, α. Below a critical temperature of 900K, and within a certain composition range, α separates into a two phase region between the α_1 and α_2 solid solutions. Figure 2.4(b) shows how the repulsive interactions between A and B atoms lead to Gibbs energy curves with two minima and a central hump below 900 K. This causes the formation of two solid solutions α_1 and α_2 below 900 K while above this temperature the Gibbs energy curves exhibit only smooth minima.

The previous examples represent two quite simple phase diagrams. In practice most phase diagrams are far more complex than this. Often the formation of intermediate phases is observed. These last mentioned phases include compounds of intermetallic or ceramic nature, whose formation may depend on relative atomic size, valency and electronegativity. In some cases, the compositions of intermediate phases correspond to an ideal atom ratio, $A_x B_y$ (where x and y are integers) and little or no stoichiometric variation is observed. In such cases, small deviations from the ideal stoichiometry cause the Gibbs energy to increase sharply, so that the Gibbs energy curve takes the form of a narrow U that, in the limit, may be virtually a line. There are other cases where an intermediate phase exists over a significant stoichiometric range. In such cases, the associated Gibbs energy vs. composition may have a much 'shallower' shape.

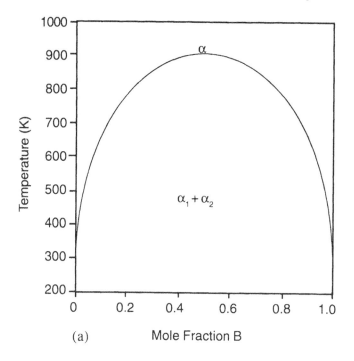

(a)   Mole Fraction B

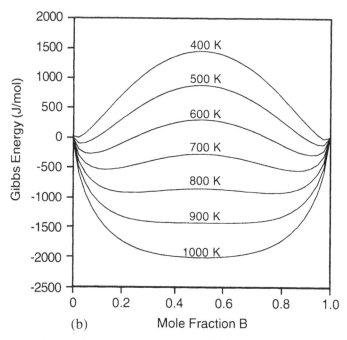

(b)   Mole Fraction B

Fig. 2.4 (a) Phase diagram for a system with a miscibility gap and (b) *G/x* curves at various temperatures. (N. Saunders and A. P. Miodownik, *CALPHAD: Calculation of Phase Diagrams.* 1996, p.50; with permission from Elsevier Science).

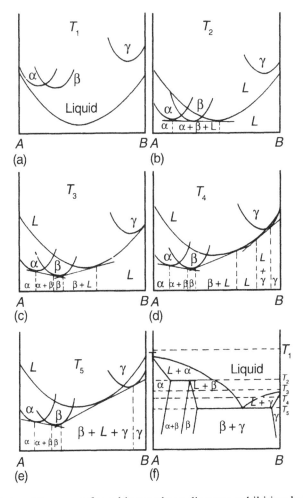

Fig. 2.5 Underlying *G/x* curves for a binary phase diagram exhibiting both peritectic and eutectic reactions. (From D. A. Porter and K. E. Easterling, *Phase Transformations in Metals and Alloys*, second edition, 1992. Chapman and Hall. Copyright Sir Alan Cottrell, used with his permission).

An example of a hypothetical binary system involving an intermediate phase is shown in Figure 2.5, illustrating the Gibbs energy vs. composition curves and the way in which peritectic and eutectic reactions arise.

2.3 CALPHAD (CALCULATION OF PHASE DIAGRAMS)

Over a period of around forty years the Calculation of Phase Diagrams for metallic and ceramic systems from thermodynamic data has become an established and vital field of materials science. Thermodynamic calculations for polymer systems have also been developed, but are not considered here. A brief introduction to some key features for metallic and ceramic systems is outlined, and some examples of real ternary systems are considered in the Case Studies, Chapter 8. The basic approach is to describe mathematically the Gibbs energy of all of the possible phases in a given system. Gibbs energy minimisation methods are then used to locate the positions of boundaries of multiphase regions.

Figure 2.4 shows Gibbs energy curves for the case of a miscibility gap. Although two phases resulted from formation of the miscibility gap, the crystal structures of both phases were the same and only one Gibbs energy curve was required to describe the two phases. However, when two or more crystallographically distinct phases are considered, two or more Gibbs energy curves are required to construct the two-phase equilibrium and the Gibbs energy at the 'end-points' of the mixing curve must then be considered.

For example, in the Cu-Ni system there are two phases, fcc and liquid, which form continuous solutions between Cu and Ni. To describe the Gibbs energy of this system the Gibbs energy of Cu and Ni must be described in both the liquid and solid state. This means that reference states for the liquid and fcc phases in pure Cu and Ni are required as well as the mixing energies of the liquid and fcc phases. The Gibbs energy is then described by the general formula

$$G = G_{ref} + G_{mix}^{xs} + G_{mix}^{ideal} \tag{14}$$

The additional term G_{ref} is a simple linear extrapolation of the Gibbs energies of the fcc and liquid phases using the equations

$$G_{ref}^{fcc} = x_{Cu} G_{Cu}^{fcc} + x_{Ni} G_{Ni}^{fcc} \tag{15}$$

and

$$G_{ref}^{liq} = x_{Cu} G_{Cu}^{liq} + x_{Ni} G_{Ni}^{liq} \tag{16}$$

where $G_{Cu,Ni}^{fcc}$ and $G_{Cu,Ni}^{liq}$ are the Gibbs energies of the fcc and liquid phases in pure Cu and Ni respectively. More generally, the reference and ideal mixing terms may be written as

$$G_{ref} = \sum_i x_i G_i^o \tag{17}$$

and

$$G_{mix}^{ideal} = RT \sum_i x_i \log_e x_i \tag{18}$$

where x_i is the mole fraction of component i and G_i^o is the Gibbs energy of phase in question at pure i.

Figure 2.6(a) shows the G/x diagram for Ni–Cu at 1523K with the fcc Ni and Cu taken as the reference states. It can be seen that Ni-rich alloys are more stable in

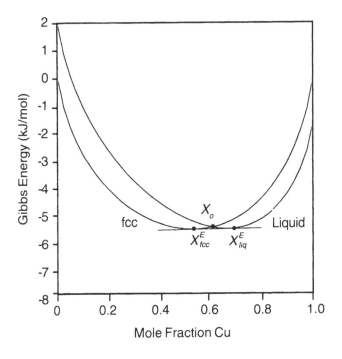

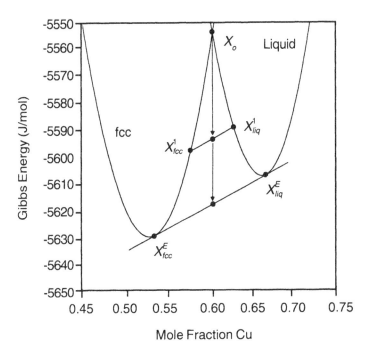

Fig. 2.6 (a) and (b) (enlargement of part of (a)) G/x curves for the liquid and fcc phases in the Ni–Cu system at 1523 (K. N. Saunders and A. P. Miodownik, *CALPHAD: Calculation of Phase Diagrams.* 1996, p.54; with permission from Elsevier Science).

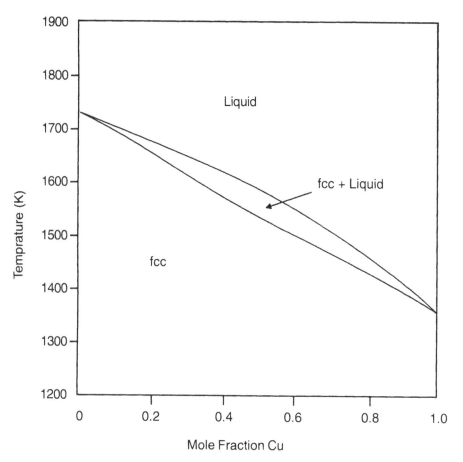

Fig. 2.7 Phase diagram for Ni–Cu showing the two-phase field between the liquid and fcc solutions. (N. Saunders and A. P. Miodownik, *CALPHAD: Calculation of Phase Diagrams.* 1996, p.55; with permission from Elsevier Science).

the fcc phase while Cu-rich alloys are more stable in the liquid phase. At x_0, the Gibbs energy of fcc and liquid phases are the same. It can then be seen that an alloy of this composition can lower its Gibbs energy by forming a two-phase mixture of fcc and liquid, with compositions x^1_{fcc} and x^1_{liq} respectively (Figure 2.6b). Similarly to the case of a miscibility gap, the Gibbs energy of alloy x_0 can be further minimised until the compositions of the fcc and liquid phases reach x^E_{fcc} and x^E_{liq} respectively at which point the phases are in equilibrium. This process is then performed over a range of temperatures and the phase diagram plotted as shown in Figure 2.7.

The corollary is that, if the thermodynamic properties of the liquid and fcc phases can be determined in sufficient detail, it would be possible to produce a phase diagram without the need for experimental determination of the phase diagram itself. Indeed this was done more than 25 years ago and the diagram was in rather good agreement with that observed experimentally (Pascoe and Mackowiak).[14a,b]

2.4 TOPOLOGICAL FEATURES OF PHASE DIAGRAMS CALCULATED USING REGULAR SOLUTION THEORY

It is instructive to look at how changes in phase diagrams can be produced by systematically varying the regular solution parameter of the Gibbs excess energy of mixing as defined by eq.(12). Such a systematic approach has been previously demonstrated by Pelton[15] and Saunders and Miodownik[13] and it is worthwhile discussing the changes and types of diagram in more detail. Figure 8 shows a series of calculated phase diagrams based on an A-B system involving two phases, one a solid the other liquid.

The first row shows the case where large repulsive interactions exist in the solid phase, producing a large miscibility gap, and the liquid regular solution parameter (Ω^{liq}) has been varied between -20 and +30 kJmol^{-1}. The first diagram with $\Omega^{liq} =$ -20 kJ mol^{-1} is characterised by a deep eutectic trough and would be typical of a system such as Au–Si. Such systems are often useful for brazing or soldering applications and can also be good glass formers. As Ω becomes less negative the general topology of the diagram remains similar; even for the case where $\Omega^{liq} = +10$ kJ mol^{-1} the system is characterised by a eutectic, although it is quite shallow. Such behaviour is typical of systems such as Ag–Bi and Ag–Pb or Cd–Zn and Cd–Pb where a retrograde solidus may occur. This is a special case where the solubility of a solid phase in equilibrium with the liquid can decrease with falling temperature. Certain alloys can then completely solidify and partially melt again as the temperature is further lowered.

As a critical value of Ω^{liq} is approached, the liquid forms its own miscibility gap and the diagram then exhibits two forms of liquid invariant reaction. The lower temperature reaction is either eutectic or peritectic while the higher temperature reaction becomes monotectic. Examples of such systems are Cu–Pb and Cu–Tl. When Ω^{liq} becomes even larger the top of liquid miscibility gap rises above scale of the graph and there is little solubility of either element in the liquid. Such a diagram is typical of Mg systems such as Mg–Fe or Mg–Mn.

The second longitudinal series shows what happens when the regular solution parameter of the solid is made less positive ($\Omega^{sol} = +15$ kJ mol^{-1}). Initially, when Ω^{liq} is negative, there is little apparent difference from the first series. There is a greater extent of solid solubility, but otherwise the type of diagram is very similar. However as Ω^{liq} becomes positive the reaction type changes to become peritectic in nature and is typical of such systems as Ag–Pt and Co–Cu. When Ω^{liq} reaches +20 kJ mol^{-1} the solid miscibility gap becomes apparent and the shape of the liquidus and solidus looks rather strange, with almost a rectangular appearance. Such a diagram is typical of systems

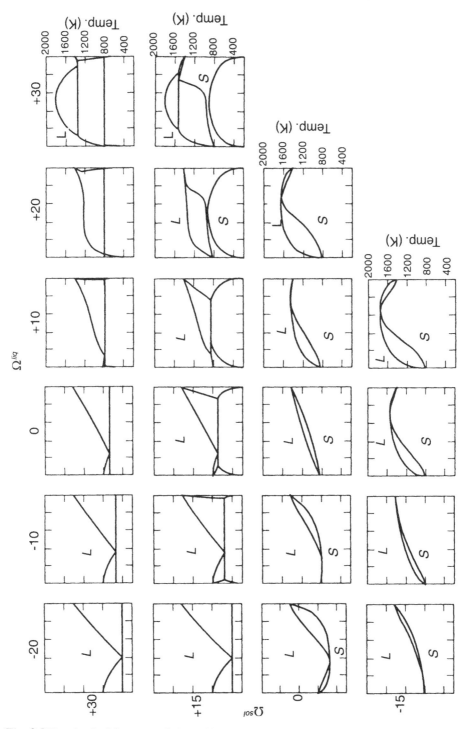

Fig. 2.8 Topological features of phase diagrams using regular solution theory. (N. Saunders and A. P. Miodownik, *CALPHAD: Calculation of Phase Diagrams*. 1996, p.56; with permission from Elsevier Science).

such as Au–Pt, Cu–Rh or Ti–W. As Ω^{liq} is further increased to $+ 30$ kJ mol^{-1}, the liquid exhibits a miscibility gap and an invariant reaction $L_1 + L_2 \rightleftharpoons$ solid occurs. This is referred to as syntectic after Rhines.[6]

The third and fourth longitudinal series takes the solid interactions to be either ideal, i.e. $\Omega^{sol} = 0$, or negative, $\Omega^{sol} = -15$ kJ mol^{-1}. The form of the diagrams is now completely changed, and they are characterised by complete solubility in the solid phase. The main features to note are

i. A minimum in the liquidus when $\Omega^{liq} < \Omega^{sol}$

ii. A maximum when $\Omega^{liq} > \Omega^{sol}$ and

iii. A smooth liquidus and solidus when $\Omega^{liq} \approx \Omega^{sol}$

The above series of calculations helps demonstrate that all types of topology of phase diagrams involving simple liquid and solid solutions can be calculated within the same simple framework. Diagrams with increasing complexity, i.e. increasing number of phases, compounds, allotropic changes in the elements etc. can also be routinely handled.

2.5 FEATURES OF PHASE DIAGRAMS IN BINARY POLYMERIC SYSTEMS

In polymer blends, consisting of intermingled chains, movement of a unit of a chain with respect to a unit of another chain is hindered by connectivity with its own chain and vice versa. Thus, the entropy of mixing term is related to molecules, rather than to atoms as in metals. Consequently, the entropy term is much lower than in metals and the Gibbs energy term is dominated by the heat of mixing term. In most polymer systems, the heat of mixing is positive and phase diagrams tend to exhibit immiscibility over most of the composition range, even in the molten state. It can be argued that in polymers, as in other types of phase diagram, there must always be some terminal solid solubility, but in high molecular weight polymers such solubility will be very small indeed.

Phase separation is an important concept, which, as displayed in a polymeric phase diagram, can parallel that of a liquid or solid miscibility gap in a metallic or ceramic system. However, there is an important exception. Figure 2.9 illustrates schematically two types of situation where there is a single phase liquid region and a miscibility gap in a binary system. One situation is where the miscibility gap lies below the single phase region and phase separation occurs on cooling from the single phase region. This situation is quite analogous to the type of behaviour seen in metallic and ceramic systems and the apex of the miscibility gap is termed the upper critical solution temperature (UCST). However, a further situation can arise in polymeric systems where the miscibility gap exists at a temperature above a single phase region. In this case, the apex of the boundary of the two phase region is designated as the lower critical solution temperature (LCST).

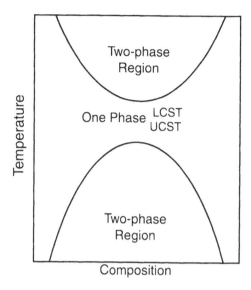

Fig. 2.9 Illustration of two types of miscibility gaps encountered in polymer blends, showing respectively an upper critical solution temperature(UCST) and a lower critical solution temperature (LCST). (After *Introduction to Physical Polymer Science*, second edition. L. H. Sperling, Copyright 1992. John Wiley and Sons Inc. Reprinted by permission of John Wiley and Sons Inc).

Concerning the LCST type, there can be a specific attractive bonding between groups of chemically different neighbouring chains even in the amorphous state, melt or glass. The favourable *A-B* interactions involve rather particular alignment of two bonding groups on the neighbouring, yet different molecules. In some respects, such bonds resemble those that would stabilise an intermetallic compound, although in this case without the lattice. Heating can then lead to a breaking of these particular bonds, a sort of local 'melting process', with the result that a single solution phase, established by negative ΔH; then phase separation occurs on heating. This type of situation is associated particularly with high molecular weight blends; a specific example is the system of polystyrene and polyvinyl methylene.

When one or two components of a ternary phase system are low molecular weight polymers, or an organic solvent is involved, then the entropy of mixing term has greater significance. Features can then emerge similar those encountered in metallic or ceramic systems. For example, a polymer system with an UCST, is represented by part of the polystyrene-cyclohexane system. (Figure 2.10). This system also illustrates the important feature of the dependence of the critical temperature on the molecular weight of the polystyrene; there is an increase in the UCST with increase in molecular weight. A further feature of this particular system is that, as the temperature is raised sufficiently above the UCST, the association between the polymer chains and the

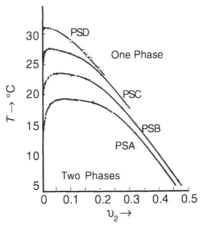

Fig. 2.10 Phase diagram for polystyrene fractions, v_2, in cyclohexane; experimentally determined. Molecular weights (viscosity average) are PSA 43,600: PSB 89,000: PSC 250,000: PSD 1,270,000 g mol^{-1}. (A. Shulz and P. J. Flory, *Journal of American Chemical Society*, 1952, **74**, 4760. Copyright, 1952, Reprinted by permission of American Chemical Society).

solvent molecules is lost. The chains then aggregate, leading again to a region of phase separation, in this case with a LCST.

Liquid crystalline polymers, whose structures show long range orientational order in one dimension, i.e. rods , constitute an important type of material. Lyotropic systems consist of polymer-solvent combinations which show a liquid crystalline phase in particular ranges of temperature and composition, and the phase diagrams show features analagous to those encountered in metallic systems. Figure 2.11 represents such a system in which the solvent lowers the melting point of the polymer and includes a narrow near-vertical two phase region known as the 'Flory chimney', which separates the regions of isotropic and liquid crystalline phases. The phase diagram further contains a region in which two liquid crystalline phases of different compositions coexist, and this region shows an UCST. This type of situation, which is associated with a three-phase reaction involving an isotropic solution and two liquid crystalline phases, may be compared with a monotectic reaction in certain binary metallic systems involving two liquid phases and a solid metallic phase. Another example, (Figure 2.12) which may be considered to be a schematic version of the important industrial system poly paraphenylene terephthalate (PPTA) and sulphuric acid as the solvent, (used in the production of Kevlar). In this particular system, the polymer is crystalline up to its melting point. The phase diagram also shows a solid crytallosolvate phase that contains both polymer and solvent molecules in its lattice, whose presence introduces two three-phase reactions. One of these is peritectic in nature where, on heating, the crystallosolvate phase decomposes into a

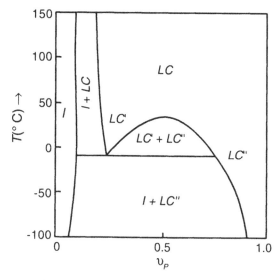

Fig. 2.11 Phase diagram illustrating the presence of an isotropic solution (I) and a lyotropic liquid crystalline phase (LC) which shows a miscibility gap in a system consisting of a rigid chain polymer and a solvent; v_p is the volume fraction of polymer.[18] (Reprinted from A. H. Windle, *A Metallurgist's Guide to Polymers: Physical Metallurgy*, 1996. R. W. Cahn and P. Haasen, eds., p.2688, with permission from Elsevier Science).

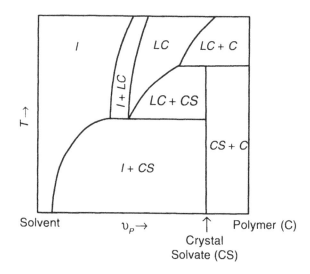

Fig. 2.12 Phase diagram representing a lyotropic polymer–solvent system which contains a crystalline polymer phase (pure polymer (C)) and a crystallosolvate phase (CS).[18] (Reprinted from A. H. Windle, *A Metallurgist's Guide to Polymers. Physical Metallurgy*, 1996. R. W. Cahn and P. Haasen, eds., p.2689, with permission from Elsevier Science).

mixture of liquid crystalline phase and crystalline polymer (cf. the situation associated with an intermetallic compound shown in Figure 6.12). In the other reaction the liquid crystalline phase decomposes on cooling into the crystallosolvate phase + solvent liquid. This may be compared with a eutectic reaction in metallic systems in which a liquid decomposes into two solid phases. In the production of Kevlar fibres, processing involves spinning from the lyotropic region and then subsequently removing of the solvent which causes crystallisation.

3. Systems Containing Two Phases

The system depicted in the view of the space model shown in Figure 3.1 consists of three components A, B and C that are completely soluble in one another in both the liquid and the solid state. The liquidus and the solidus, which are represented by curved surfaces, enclose a space constituting the two-phase region of liquid + solid solution. This two-phase region may be considered as consisting of a bundle of tie-lines of varying direction; each tie-line links the composition points of liquid and solid co-existing at a particular temperature, and no two-tie-lines at the same temperature can cross one another. Experiments or thermodynamic calculations are required to establish the position and the direction of the tie-lines in any particular system.

The determination of liquidus and solidus temperatures for a series of alloys in a system provides data for plotting isotherms on the liquidus and solidus surfaces. Such isotherms may be conveniently depicted on a projected view on the base of the space model, e.g. Figure 3.2 show a liquidus projection or 'liquidus plan' for the Ag–Au–Pd system.

It should be noted that in some systems the liquidus and solidus surfaces may exhibit a maximum or a minimum point at which the liquidus and solidus touch one another.

The process of equilibrium solidification in the system shown in Figures. 3.1a and b may be illustrated by reference to an alloy of composition indicated by point X. On cooling through the liquidus temperature to temperature T_1, solidification begins by the formation of crystals of solid solution, which contain more of component B than the original liquid. The compositions of the liquid and solid are given by the extremities of the tie-line L_1S_1. The positions of several tie-lines at lower temperatures are also shown and each tie-line passes through the alloy composition point. The extremities of these tie-lines lie on the liquidus and solidus surfaces respectively, and the paths $L_1L_2L_3L_4$ and $S_1S_2S_3S_4$ represent the progressive change of composition of liquid and solid during solidification. At any temperature application of the lever rule to the appropriate tie-line enables the relative amounts of liquid and solid to be calculated, e.g. at T_2 (Figure 3.3a).

$$\frac{\text{Weight of Liquid}}{\text{Weight of Solid}} = \frac{X_2S_2}{X_2L_2}$$

$$\%\text{Liquid} = \frac{X_2S_2}{L_2S_2} \times 100$$

$$\%\text{Solid} = \frac{X_2L_2}{L_2S_2} \times 100$$

With falling temperature, the percentage of liquid decreases until at the solidus temperature, T_4, the composition of the solid solution attains the alloy composition and

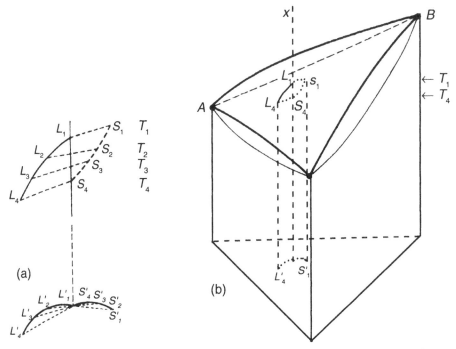

Fig. 3.1 (a) and (b), Space model of system showing complete solid solubility of the components; the diagrams also depict the course of solidification of a typical alloy X.

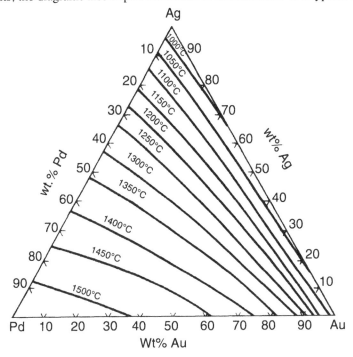

Fig. 3.2 Liquidus projection for the Ag–Au–Pd system.(From Metals Handbook, 1948, Reproduced with permission from ASM International, Materials Park, OH 44073-0002, U.S.A).

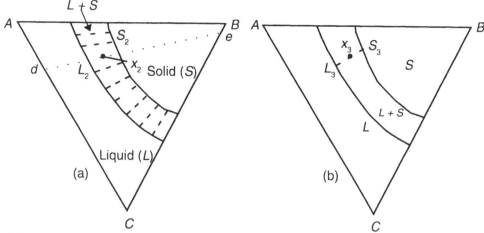

Fig. 3.3 (a) and (b). Isothermal sections through the space model of Figure 3.1 for temperatures T_2 and T_3 respectively.

no liquid remains. The case of non-equilibrium freezing, leading to coring, and the resultant extension of the temperature range of solidification has been described by Rhines.[6]

As stated above, if it is required to know the actual compositions of liquid and solid co-existing at any temperature, experimental work or thermodynamic calculations are necessary. For a given alloy, the tie-lines at various temperatures do not generally lie in the same vertical plane and the paths of the extremities of the tie-lines on the liquidus and solidus are generally curved. This effect is also illustrated in Figures 3.1a and b by the projected view on the base of the space model of paths representing the change in composition of liquid and solid. The tie-line L_1S_1 represents the beginning of solidification, and L_4S_4 the end of solidification.

Tie-lines for alloy X are also indicated in the isothermal sections through the space model at temperatures T_2 and T_3 respectively (Figures 3.3a and b). Other typical tie-lines for the liquid + solid region are included in the sections, and it is only by coincidence that a tie-line may point directly towards a corner of the system.

The vertical section de shown in Figure 3.4 is taken along the direction of the tie-line L_2S_2 at temperature T_2. At other temperatures within the liquidus solidus range, the tie-lines intersect the plane of the section at the composition point of the alloy but the extremities of the tie-lines do not lie in the plane. Thus a vertical section of this type through a ternary system differs essentially from a binary phase diagram since, in general, the tie-lines do not lie in the plane of the section. During the solidification of alloys whose compositions lie in the section, the liquid and solid compositions change out of the plane of the section. On such a section it is not therefore correct to assume that a horizontal line cutting the liquidus and solidus surfaces in this plane necessarily defines the compositions of the liquid and solid co-existing in equilibrium.

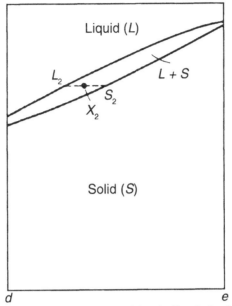

Fig. 3.4 Vertical section taken along direction of the tie-line L_2S_2 of Figure 3.3(a).

Table 3.1

Composition (wt.%)			Temperature, °C	
A	B	C	Liquidus	Solidus
50	50	-	975	950
50	-	50	920	850
-	50	50	840	800

3.1 PROBLEM

The melting points of A, B and C are 1000°C, 900°C and 750°C, respectively. The constituent binary systems of the ternary system ABC each show complete solubility in both the liquid and the solid states.

The data shown in Table 3.1 refer to three binary alloys:

For the ternary system draw and label diagrams consistent with the above data to represent the possible forms of

a. a liquidus projection, showing isotherms for 950°C and 850°C.

b. a solidus projection showing isotherms for 950°C and 850°C.

c. an isothermal section at 950°C.

d. a vertical section joining A to the mid-point of BC. Assume that the liquidus and solidus surfaces do not exhibit a maximum or a minimum.

4. Systems Containing Three Phases

4.1 INTRODUCTION

The following cases, a–e, relating to metallic and ceramic systems will be considered:

a. Liquid (L) + two primary solid solutions (α and β) giving a three phase eutectic reaction $L \rightarrow \alpha + \beta$;

b. Liquid (L) + two primary solid solutions (α and β) giving a three-phase peritectic reaction $L + \alpha \rightarrow \beta$;

c. Transition between eutectic and peritectic reactions;

d. Two liquids (L_1 and L_2) + one primary solid solution (α) giving a three phase monotectic reaction $L_1 \rightarrow L_2 + \alpha$;

e. Systems containing a solid state miscibility gap that closes within the system;

f. Miscibility gaps in ternary polymeric systems.

4.2 SYSTEMS CONTAINING A EUTECTIC REACTION $L \rightarrow \alpha + \beta$

In the hypothetical system ABC shown in Figure 4.1 it is assumed that there is complete liquid miscibility. Binary system AC shows complete solid solubility, while systems AB and BC show partial solubility and are of the eutectic type. The solid solutions are termed α and β respectively, the former extending completely across the AC side of the system; in the ternary system the single-phase regions of α and β are separated from each other by a two-phase $\alpha + \beta$ region. The temperature of the eutectic point M representing the reaction $L \rightarrow \alpha + \beta$ in system AB is assumed to be higher than that of point N representing the reaction $L \rightarrow \alpha + \beta$ in system BC.

The view of the space model (Figure 4.1) shows that the liquidus is made up of two surfaces $AMNC$ and BMN, corresponding to the primary solidification of α and β respectively. These surfaces intersect along a eutectic curve or 'valley' MN. In principle, this curve MN may show a maximum or minimum point, but it is assumed in the example illustrated that the curve falls smoothly as it traverses the system from side AB to side BC. DG and EF are curves joining the points representing the respective compositions of the α and β phases formed in the eutectic reactions in the binary systems (these reactions are represented by the horizontal straight lines DME and GNF). Surfaces $ADGC$ and BEF form part of the solidus of the system, and each of the liquid + solid regions (i.e. liquid + α and liquid + β) enclosed by the respective liquidus and solidus surfaces resembles the two-phase region of the solid solution system previously discussed.

35

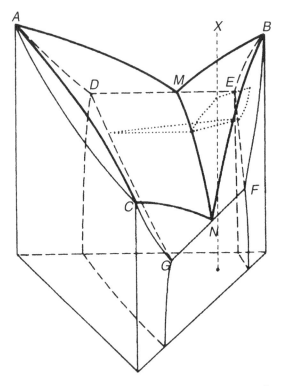

Fig. 4.1 Space model of system showing a eutectic reaction $L \rightarrow \alpha + \beta$.

The three curves MN, DG, and EF do not lie in the same plane; curve MN lies above the surface $DEFG$, so that there are three curved surfaces, namely $DMNG$, $MEFN$, and $DEFG$, which enclose a three-phase space, representing the co-existence of liquid, α and β. Each of these surfaces is made up of tie-lines, representing the co-existence of L and α, L and β, and α and β respectively. Surfaces $DMNG$, $MEFN$ separate the three-phase space from the $L + \alpha$ and $L + \beta$ regions respectively: $DEFG$ separates the three-phase space from the $\alpha + \beta$ region, and is thus the solidus surface for the $\alpha + \beta$ portion of the system. (The form of a vertical section that cuts through the three-phase space is shown in Figure 4.8). Where the three-phase space terminates in the binary systems AB and BC it shrinks to the horizontal straight lines DME and GNF respectively. (In a later section (page 56) reference will be made to a ternary system in which the three-phase space closes within the system).

4.2.1 THE EUTECTIC REACTION

The three-phase space is associated with the eutectic reaction $L \rightarrow \alpha + \beta$ which occurs over a range of temperature in a ternary system, not at a constant temperature as

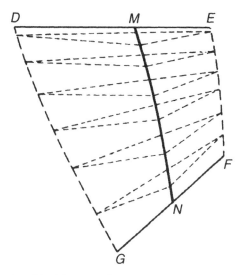

Fig. 4.2 Three-phase region (liquid, α and β) of Figure 4.1.

in a binary system. The three-phase region may be considered as being made up of a stack of isothermal tie-triangles, and its form is illustrated in Figure 4.2, which shows typical tie-triangles at several temperatures. The corners of these triangles represent the compositions of the phases co-existing in equilibrium at the particular temperatures concerned, and as the eutectic reaction proceeds with falling temperature the liquid composition changes along curve MN, and the compositions of α and β change along DG and EF respectively. Thus, while three phases co-exist, the composition of each phase is restricted to a curve as the temperature changes.

The process of eutectic solidification under equilibrium conditions may be illustrated by reference to the alloy X shown in Figure 4.1; this alloy, when solid, consists of primary crystals of β plus α eutectic mixture of α + β. During solidification, the primary stage (i.e. the separation of β) is the same in principle as the solidification of an alloy in a system showing complete solid solubility. Thus, the composition of the liquid changes along a path on the liquidus surface as the temperature falls, and the β-composition changes on a path on the solidus. Then at a certain temperature T_1, before solidification is completed, the liquid composition reaches a point on curve MN and simultaneously the β composition reaches a point on curve EF. The situation at T_1 is illustrated by the tie-triangle in Figure 4.1 and in the projected view of the tie-triangle shown in Figure 4.3a; the alloy composition point is seen to lie on the Lβ tie-line that forms one of the boundaries of the Lαβ triangle at this temperature. The relative amounts of L and β may be obtained by applying the lever principle to the tie-line. With a fall in temperature to T_2, the three-phase region is entered (i.e. liquid, α and β co-exist), and the alloy composition point now lies inside the tie-triangle (Figure 4.3b). The compositions of the liquid, α and β are given by the points. L_2, α_2, and β_2 respectively, which are the corner points of the triangle, and application of the 'centre-of-gravity' principle gives the amounts of the phases present.

$$\% \text{Liquid}\,(L_2) \quad = \quad \frac{xL'_2}{L_2L'_2} \times 100$$

$$\% \beta_2 \quad = \quad \frac{x\beta'_2}{\beta_2\beta'_2} \times 100$$

$$\% \alpha_2 \quad = \quad \frac{x\alpha'_2}{\alpha_2\alpha'_2} \times 100$$

In the present case it is assumed that the tie-triangles have approximately the same size and shape throughout the freezing range of the alloy X, but in actual systems there can be appreciable differences in these characteristics of the triangles at the various temperatures.

During cooling to some lower temperature, T_3, the compositions of liquid, α and β change along the curves MN, DG, and EF respectively. The tie-triangle corresponding to temperature T_3 is shown in Figure 4.3c. The position of the triangle relative to the alloy composition point X is different from that at T_2, and application of the 'centre-of-gravity' principle shows that the amount of liquid is less at T_3 than at T_2, while the amounts of α and β are greater; this fact is in accordance with the eutectic nature of the reaction, $L \rightarrow \alpha + \beta$.

Compositional changes in the liquid, α and β continue during further cooling until at the solidus temperature, T_4, the eutectic reaction is completed. The situation is depicted by the tie-triangle in Figure 4.3d, where the alloy composition point is now seen to lie on the $\alpha + \beta$ side of the tie-triangle, that is to say, no liquid remains. The relative amounts of α and β may be obtained by applying the lever principle to the $\alpha\beta$ tie-line.

The process of eutectic solidification as described, may be viewed in terms of the displacement of the tie-triangles, relative to the alloy composition point, solidification being completed when the $\alpha\beta$ side of a triangle contains the alloy composition point.

4.2.2 PROJECTED VIEW OF THE SYSTEM

The solidification process of alloys in a given system may be traced, using data both for the liquidus of the system and for the solid state regions. The data may be conveniently presented as a projected view of the system on the base of the space model. Figure 4.4 shows the projected view of the liquidus, with curve mn representing the eutectic valley MN; the figure also includes the solidus projection, showing dg and ef representing the solidus curves DG and EF which define the limits of the α and β solid solution regions and of the $\alpha + \beta$ region at the completion of solidification. The α and β liquidus surfaces are represented by the regions $AmnC$ and Bmn respectively, and the corresponding α, β, and $\alpha + \beta$ solidus surfaces by $AdgC$, Bef, and $defg$ respectively.

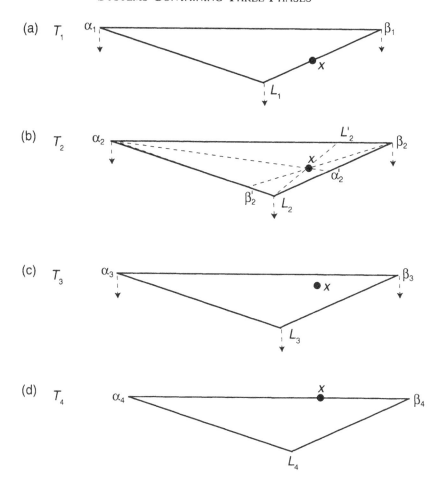

Fig. 4.3 Tie-triangles at various temperatures illustrating the progress of the eutectic reaction in alloy X of Figure 4.1.

If there are no solubility changes after the completion of solidification, the positions of the boundaries of the solid state regions in the projected view (Figure 4.4) are the same as those shown in an isothermal section through the solid state portion of the system (Figure 4.7b). In practice, changes of solid solubility, after solidification is completed, will generally cause the positions of the boundaries to differ in the two types of diagram, but it is sometimes convenient for purposes of discussion to ignore these changes.

The position of a typical tie-triangle is also shown in Figure 4.4, but it should be appreciated that whereas a tie-triangle is isothermal in nature, the curve *mn*, *dg*, and *ef* are projections of curves extending over a range of temperature, as indicated by the arrows.

The projected view of the system may be used as follows to trace the solidification processes of various alloys. Whether the location of the alloy composition lies either in

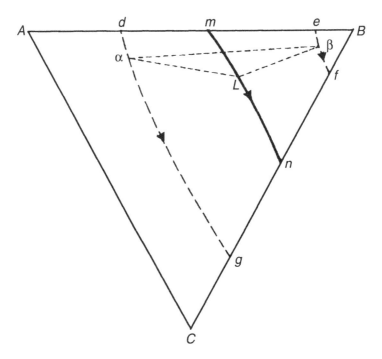

Fig. 4.4 Projected view on the base of the space model of the system shown in Figure 4.1, including a typical tie-triangle.

region *AmnC* or in region *Bmn* defines whether α or β will be the primary phase during solidification; if the alloy composition lies on curve mn then solidification occurs directly by the eutectic reaction without the separation of primary α or β. Without relevant data for a given system it is not possible to deduce from the diagram the composition of the primary phase at the beginning of solidification, or the precise changes in composition of the liquid and solid during the primary stage of solidification. However, if the primary β solid solubility is very limited the resultant changes in the liquid composition may be estimated by the procedure discussed later (see page 67).

If the alloy composition lies within one of the single-phase regions *AdgC* or *Bef*, then solidification is completed in the manner described for a system showing complete solubility. For alloys lying in the two-phase region *defg*, the liquid composition reaches curve *mn* during solidification and with further cooling, the eutectic reaction proceeds until the liquid is consumed.

The possibility of the eutectic curve showing a maximum or minimum has already been mentioned. In a projected view these cases are distinguished by the directions of the arrows on the projection of the eutectic curve (e.g. as in Figures. 4.5 and 4.6).

At the maximum or minimum points, the tie-triangles reduce to a tie-line joining the compositions of *L*, α, and β in equilibrium. Detailed diagrams are given by Rhines.[6]

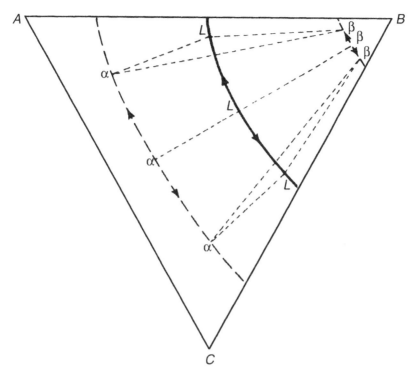

Fig. 4.5 Projected views of system in which the eutectic valley shows a maximum.

4.2.3 ISOTHERMAL SECTIONS

A representative isothermal section through the space model shown in Figure 4.1 at a temperature between that of the eutectics in systems *AB* and *BC* is shown in Figure 4.7a. The triangular section (i.e. tie-triangle) through the $L + \alpha + \beta$ region is seen in relation to the surrounding single-phase and two-phase fields. Typical tie-lines are shown traversing the two-phase regions.

Figure 4.7b shows an isothermal section in the solid state at a temperature just below that of eutectic *BC*, and, as previously mentioned, if solid solubility changes after the completion of solidification are ignored, the positions of the boundary curves are identical with those in the solidus projection of Figure 4.4.

4.2.4 VERTICAL SECTION

The form of a vertical section passing through corner *B* and traversing the system to the midpoint *r* of side *AC* is shown in Figure 4.8. The occurrence of the eutectic reaction over a range of temperatures is indicated by the three-phase region, $L + \alpha + \beta$. It should

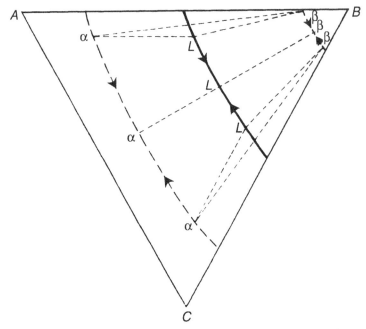

Fig. 4.6 Projected view of system in which the eutectic valley shows a minimum.

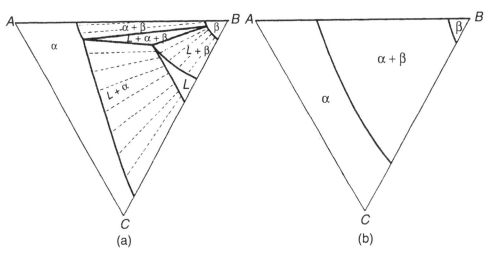

Fig. 4.7 Representative isothermal sections through the space model of Figure 4.1(a) at a temperature between that of the eutectics in systems *AB* and *BC* (b) at a temperature below that of eutectic *BC*.

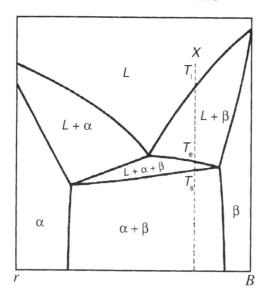

Fig. 4.8 Vertical section of system shown in Figure 4.1 from corner *B* to the mid-point *r* of side *AC*.

be noted that the boundaries of this region do not necessarily appear as straight lines, since the boundaries are sections through surfaces formed by series of tie-lines which do not generally lie in the plane of the section; further, in the two-phase regions, the tie-lines are not generally in the plane of the section.

While this type of section does not provide information on the compositions of phases co-existing in equilibrium, it does show the temperatures at which phase changes occur in given alloys, and also the stages of solidification, e.g. for alloy X, T_1, is the liquidus, T_e, is the temperature at which the eutectic crystallization begins, and T_s, is the temperature at which solidification is completed.

4.3 SYSTEMS CONTAINING A PERITECTIC REACTION $L + \alpha \rightarrow \beta$

The hypothetical system *ABC* depicted in Figure 4.9 represents complete liquid miscibility; the binary system *BC* shows complete solid solubility, while systems *AB* and *AC* involve partial solid solubility, and are of the peritectic type. The solid solutions are termed α and β respectively, of which the latter extends completely across the *BC* side of the system. The temperature of the peritectic reaction $L + \alpha \rightarrow \beta$ in system *AB* is greater than that of the same reaction in system *AC*.

The liquidus is seen to consist of two surfaces, *AMN* and *MBCN*, corresponding to the primary separation of α and β respectively, during solidification. These surfaces intersect along the peritectic curve *MN*, which is not a valley as in the eutectic type of system, but is rather like a 'shoulder'. The curves *EF* and *DG* join the points representing

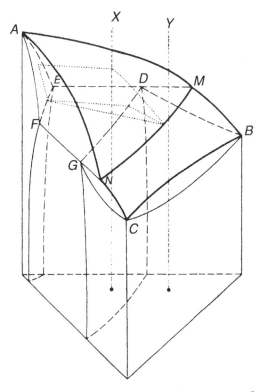

Fig. 4.9 Space model of system showing a peritectic reaction $L + \alpha \rightarrow \beta$.

the respective compositions of α and β taking part in the peritectic reaction in the binary systems (these reactions are represented by the horizontal straight lines *EDM* and *FGN*). Surface *AEF* is the solidus of the liquid + α region, and *DBCG* is the solidus surface of the $L + \beta$ region. The curves *MN*, *DG*, and *EF* do not lie in the same plane, and curve *DG* lies below the surface *EMNF*. The three-phase space, representing the co-existence of liquid, α and β taking part in the peritectic reaction, is bounded by the three surfaces *EMNF*, *DMNG*, and *EDGF*, the last mentioned being the solidus for the $\alpha + \beta$ part of the system; each of these surfaces is made up of tie-lines, namely $L\alpha$, $L\beta$, and $\alpha\beta$ tie-lines respectively. A vertical section through the three-phase space is shown in Figure 4.16. Where the space terminates in the binary systems, it shrinks to become a line - either line *EDM* or *FGN*. (The nature of a system in which the three-phase space closes within the system is discussed later, page 56).

4.3.1 THE PERITECTIC REACTION

Figure 4.10 illustrates the form of the three-phase space and shows some of the representative tie-triangles that constitute the space. As the peritectic reaction proceeds

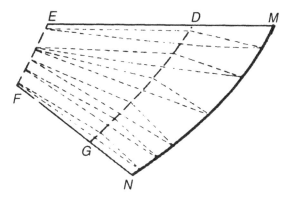

Fig. 4.10 Three-phase region (liquid, α and β) of Figure 4.9.

with falling temperature, the liquid composition changes along curve *MN*, and the compositions of α and β change along *EF* and *DG* respectively.

The process of peritectic solidification may be illustrated by reference to alloys lying in the region *EMNF*. These alloys commence solidification by the separation of crystals of α-phase, as, for example, in alloy *X* of Figure 4.9; the liquid composition changes along a curved path on the liquidus surface, in the typical manner previously described, while the composition of α changes along the α-solidus surface. At some temperature, say T_1, the liquid composition reaches a point on the peritectic curve *MN* and simultaneously the α composition reaches a point on curve *EF*. At this temperature the *L*α tie-line forms one of the boundaries of the *L*αβ tie-triangle. Further cooling brings the alloy into the three-phase region $L + α + β$, and the peritectic reaction begins to occur. During the reaction either the liquid phase or the α phase may be consumed, depending on whether the alloy composition lies in the region *EDGF* or *DMNG* (corresponding to typical alloys such as 1 and 2 in the binary system shown in Figure 4.11); these two cases in the ternary system may be illustrated by reference to the tie-triangles in a manner analogous to that discussed for the eutectic reaction in the previous section. Consider first alloy *X* lying in the region *EDGF*, that is, an alloy which consists of α and β phases when solid. As described above, primary solidification of α occurs until temperature T_1 is reached, when the situation is as illustrated in Figure 4.12a; the alloy composition point lies on the *L*α tie-line forming one side of the *L*αβ tie-triangle, and a slight decrease in temperature to T_2 will lead to the formation of some β by the reaction $L + α \rightarrow β$. The tie-triangle at T_2 is shown in Figure 4.12b and the alloy composition point *X* is now within the triangle. At a lower temperature T_3 (Figure 4.12c), the point lies nearer to the αβ side of the triangle. The peritectic nature of the reaction, viz., $L + α \rightarrow β$ may be demonstrated by applying the 'centre of gravity' principle to the tie-triangles to calculate the percentages of the phases present at T_2 and T_3. It is found that the amount of liquid and the amount of α have decreased during cooling from T_2 to T_3, while the amount of β has increased. At

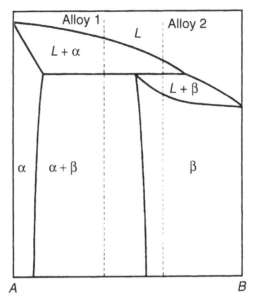

Fig. 4.11 Binary peritectic system.

T_4, the solidus temperature for the alloy, the alloy composition point lies on the $\alpha\beta$ side of the tie-triangle (Figure 4.12d), indicating that all the liquid has been consumed in the peritectic reaction.

For an alloy such as Y, lying in the region *DMNG*, primary solidification of α is followed by the peritectic reaction, but in this case the α phase is consumed during the peritectic reaction and L and β remain. The situation at the temperature at which the peritectic reaction is completed is depicted by the tie-triangle in Figure 4.13, the alloy composition point Y now lying on the $L\beta$ side of the triangle. With further cooling solidification continues by the separation of β, and the change in liquid composition occurs on a curved path along the *MBCN* portion of the liquidus; (the liquid composition is no longer restricted to movement along curve *MN*, since only one solid phase now exists in equilibrium with the liquid). Solidification is completed in the manner characteristic of solid solution alloys, and the final structure of the alloy is β solid solution.

4.3.2 PROJECTED VIEW OF THE SYSTEM

A projected view of the ternary system is shown in Figure 4.14. Curve *mn* is the projection of the liquidus curve *MN*. *Amn* is the region of primary α solidification, and *mBCn* is the region of primary β solidification. *dg* and *ef* are the projections of the solidus curves *DG* and *EF*, and the α and β solidus surfaces are indicated by the regions *Aef* and *dBCg*, which are the α and β solid solution regions at the completion of solidification. The $\alpha + \beta$ region is *edgf*. The position of a typical tie-triangle is indicated in the diagram.

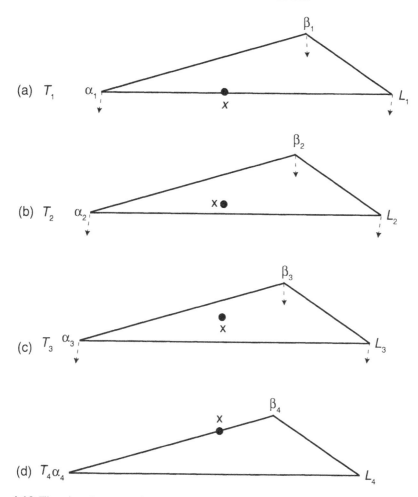

Fig. 4.12 Tie-triangles at various temperatures illustrating the progress of the peritectic reaction in an alloy, X, that is $\alpha + \beta$ when solid.

The projected view of the system may be used to trace the solidification of alloys in a manner analogous to that described for the eutectic system in the section 4.2.2. The regions representing the projection of the liquidus are used to determine which phase will separate first during solidification and the regions defined by the projections of the solidus curves dg and ef are used to determine the final solid state constitution. Alloys lying in the region $emnf$ undergo the peritectic reaction, since the liquid composition reaches a point on curve mn as a result of α separation; if the alloy is in region $edgf$, solidification is completed at the end of the peritectic, the liquid being consumed and $\alpha + \beta$ remaining; within region $dmng$, the α is consumed, and solidification is completed by the direct separation of β from liquid giving a final structure of β solid solution.

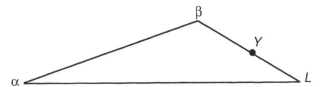

Fig. 4.13 Tie-triangle illustrating the completion of the peritectic reaction in an alloy Y, that consists only of β-phase when solidification is finally completed.

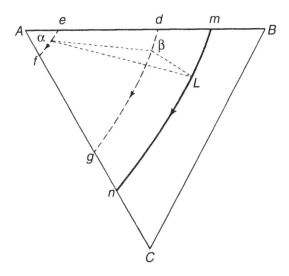

Fig. 4.14 Projected view on the base of the space model of the system shown in Figure 4.9, including a typical tie-triangle.

4.3.3 ISOTHERMAL SECTIONS

Figure 4.15a shows a representative isothermal section through the space model at a temperature between that of the peritectic reactions in systems AB and AC; the $L + \alpha + \beta$ region is seen in relation to the adjoining regions.

Figure 4.15b shows an isothermal section at a temperature below that of the peritectic reaction in system AC; if solid solubility changes after the completion of solidification are ignored, then the boundaries in the isothermal section correspond to those in the solidus projection (Figure 4.14).

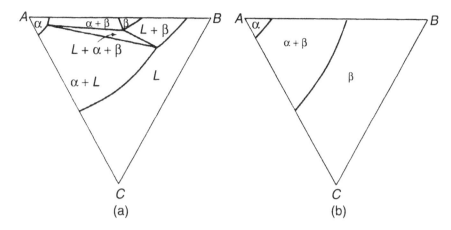

Fig. 4.15 Representative isothermal sections through the space model of Figure 4.9. (a) at a temperature between that of the peritectic reactions in systems *AB* and *AC* (b) at a temperature below that of the peritectic reaction in system *AC*.

4.3.4 Vertical Section

A vertical section passing through corner *A* to the midpoint *r* of side *BC* is shown in Figure 4.16, the occurrence of the peritectic region over a range of temperature being indicated by the extent of the *L* + α + β region. The comments made previously on vertical sections apply here also.

4.4 TRANSITION BETWEEN EUTECTIC AND PERITECTIC THREE-PHASE REACTIONS

In the preceding sections the occurrence of eutectic and peritectic reactions has been discussed on the basis of tie-triangles that define the compositions of the reacting phases at various temperatures. A criterion for distinguishing between the two types of reaction is commonly based on a consideration of the tie-triangles in relation to their displacement, with falling temperature, along the curve representing the change of liquid composition. This criterion will be briefly presented here, although it should be noted that it is not satisfactory under certain conditions of compositional changes during the reaction; the reasons for the limitations of the criterion are not described here, but have been discussed by Hillert.[21]

In Figures 4.17a and c the projected view of the liquid composition curve is shown in relation to the tie-triangle for the peritectic and eutectic cases respectively. In Figure 4.17a at the temperature considered, the tangent drawn to the liquid composition

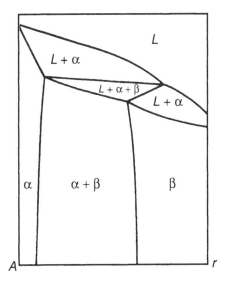

Fig. 4.16 Vertical section of system showed in Figure 4.9 from corner A to the midpoint r of side BC.

curve, at the liquid composition corner of the tie-triangle, lies outside the triangle, and therefore does not cut the $\alpha\beta$ side of the triangle; the reaction that is proceeding is taken to be peritectic in nature and with falling temperature displacement of the triangle is associated with the reaction $L + \alpha \rightarrow \beta$. Figure 4.17c illustrates the eutectic case, in which the tangent to the liquid composition curve cuts the $\alpha\beta$ side of the tie-triangle; then with decreasing temperature the liquid deposits a mixture of two solids i.e. $L \rightarrow \alpha + \beta$. Figure 4.17b depicts a situation in which, as a transition stage, the tangent to the liquid composition curve coincides with the $L\beta$ tie-line.

    This differentiation between the peritectic and eutectic examples may be compared with that made on the basis of the relative compositions of the phases involved in these types of reactions in binary systems. (Figure 4.18.)

    It is possible that within a given ternary system a transition may occur from one type of three-phase reaction to another, e.g. from peritectic to eutectic. Such an occurrence is illustrated by the ternary system shown in Figures 4.19 and 4.20, in which the direction of the liquid composition curve for the three-phase reaction changes relative to those of the α and β composition curves within the system in such a way as to lead to a transition. Binary system BC is assumed to show complete solid solubility, AB is a peritectic system $L + \alpha \rightarrow \beta$, and AC is a eutectic system $L \rightarrow \alpha + \beta$.

    The peritectic temperature in AB is higher than the eutectic temperature in AC. It should be noted that the projections of the liquid composition curve and of the β composition curve in Figure 4.20 do not indicate an actual intersection of these curves; the latter curve in fact passes under the former in the space model.

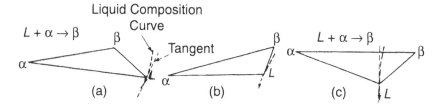

Fig. 4.17 Projected views of liquid composition curve (shown dashed) in relation to a tie-triangle for the peritectic and eutectic cases and for the transition case.

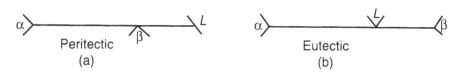

Fig. 4.18 Schematic representation of a peritectic and eutectic reaction in a binary system.

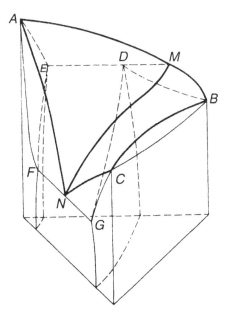

Fig. 4.19 Space model of system showing a transition between a peritectic and a eutectic reaction.

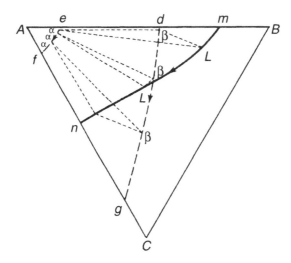

Fig. 4.20 Projected view of system of Figure 4.19.

The three-phase region in the system is composed of a continuous series of tie-triangles of various shapes, part of the region being associated with the peritectic reaction and part with the eutectic reaction. Of the alloys that undergo the three-phase reaction, some, with compositions lying towards side AB of the system, will be clearly wholly peritectic in their three-phase stage of solidification, and others lying towards AC will be wholly eutectic; there is, however, an intermediate range of compositions in which there occurs a change in the nature of the three-phase reaction during the course of solidification.[21]

If the temperature of the eutectic in system AC was higher than that of the peritectic in AB then a transition from eutectic to peritectic could occur within the ternary system.

4.5 SYSTEMS CONTAINING A MONOTECTIC REACTION $L_1 \rightarrow L_2 + \alpha$

The presence of a monotectic reaction in a system is associated with a miscibility (or solubility) gap in the liquid state.[*]

The concept of a miscibility gap may be illustrated by reference to the portion of binary diagram shown in Figure 4.21; there is partial solubility in the liquid state, and the region in which two liquidus co-exist is termed the miscibility gap. In the case shown, as the temperature increases the tie-lines spanning the miscibility gap become shorter, until, at the temperature of the critical point, c, the tie-line has zero length and the gap closes; above this temperature there is complete liquid solubility.

---

[*] The work of Wright and Thompson near the end of the 19th century (see p. 2, Ch.1) on systems such as Pb—Zn—Sn and Pb—Zn—Ag included the identification of liquid miscibility gaps, and introduced the concept of critical curves with associated tie-lines.

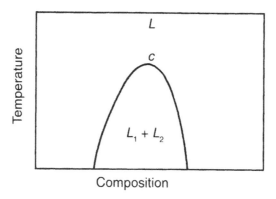

Fig. 4.21 Liquid miscibility gap in a binary system showing critical point, c.

An analogous case for a ternary system is shown in Figure 4.22a where a liquid miscibility gap in binary system AB extends into the ternary system to form a region that has the shape of a partial dome. It will be seen that the miscibility gap closes within the ternary system, and that outside the gap there is complete liquid solubility.

An isothermal section through the miscibility gap is shown in Figure 4.22b; the diagram indicates the positions of typical L_1L_2 tie-lines and of the critical point c_1 at which the tie-line reaches zero length. (Figure 4.22c shows the miscibility gap in an isothermal section of the Pb–Zn–Sn system). For each temperature considered there is a similar critical point, and these points constitute the critical curve $cc_1c_2c_3$, etc., shown in Figure 4.22a. An analogous type of critical curve exists for the case of a solid state miscibility gap that closes within a ternary system (see page 56).

The example shown is not the only possible form for a liquid miscibility gap; for example, there may be a critical point within the system at a temperature higher than the binary critical point.

A liquid miscibility gap can give rise to a monotectic type of reaction, and Figure 4.23 shows a part of a ternary system in which one of the binary systems, AB, shows the reaction

Liquid$_1$ (L_1) $\rightarrow$ Liquid$_2$ (L_2) + α solid solution

The partial dome representing the $L_1 + L_2$, region lies upon the liquidus surface of the ternary system; the critical point S lies at a lower temperature than that of the binary monotectic reaction. There exists a three-phase space representing the co-existence of L_1, L_2, and α solid solution, the space being made up of $L_1L_2\alpha$ tie-triangles, and the occurrence of the reaction $L_1 \rightarrow L_2 + \alpha$ can be considered in terms of these tie-triangles. As the reaction occurs with falling temperature, the composition of L_1 changes along RS, that of L_2 along TS and that of α along PQ. Within the ternary system the three-phase space terminates at the tie-line QS.

The solidification of a typical alloy, X, may be used to illustrate the monotectic reaction (Figures 4.23 and 4.24). Solidification begins by the separation of primary α,

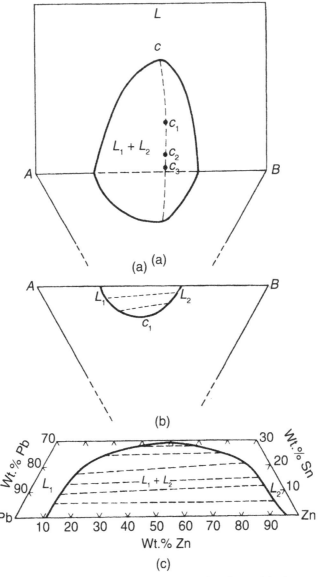

Fig. 4.22 (a) Liquid miscibility in a ternary system showing critical curve $cc_1c_2c_3$ (b) Isothermal section through the miscibility gap of Figure 4.22a showing tie-lines. Part of the isothermal section at 520°C of the Pb–Zn–Sn system showing tie-lines in the liquid miscibility gap. (From *Metals Handbook*, 1948, Reproduced with permission from *ASM International*, Materials Park, OH,44073-0002, U.S.A).

and the compositional changes that occur with falling temperature eventually bring the liquid composition on to curve *RS* and the α-composition on to curve *PQ*, at some temperature T_1; the equilibrium at T_1 is illustrated by the tie-triangle in Figure 4.25a in which the alloy composition point lies on the $L_1\alpha$ side of the triangle. At a lower

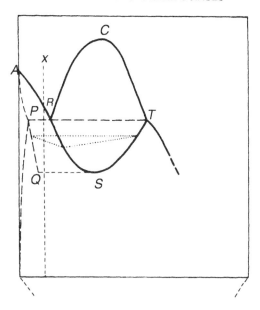

Fig. 4.23 View of part of the space model of a ternary system containing a miscibility gap giving rises to a monotectic reaction.

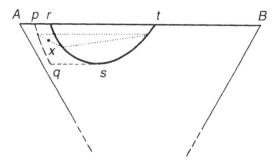

Fig. 4.24 Projected view of Figure 4.23, including a tie-triangle.

temperature, T_2 (Figure 4.25b), the alloy composition point X now lies within the tie-triangle, and three phases exist, namely α, L_1, and L_2. At a still lower temperature, T_3 (Figure 4.25c), X lies on the αL_2 side of the triangle, indicating that none of L_1 remains. Thus, at T_3, the monotectic reaction is completed for the alloy X, and on further cooling, L_2 (now no longer constrained to move along curve TS) changes composition along a curved path on the liquidus surface, depositing α solid solution. The further course of solidification depends on the nature of the remaining portion of the system, not shown in Figure 4.23.

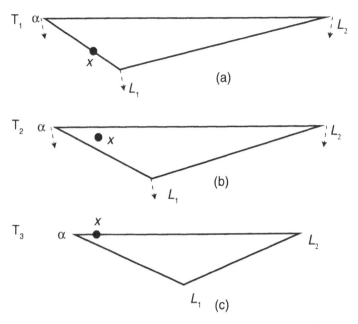

Fig. 4.25 Tie-triangle illustrating the progress of the monotectic reaction in an alloy X.

For a fuller description of the monotectic reaction, reference may be made to other accounts[2,3] where it is assumed that solid solubility is negligible, and for this condition, it is shown how either of the two liquidus (i.e. either L_1 or L_2) may be consumed during the reaction, depending on the alloy composition.

4.6 SYSTEMS CONTAINING A SOLID STATE MISCIBILITY GAP THAT CLOSES WITHIN THE SYSTEM

In the eutectic and peritectic systems previously discussed, the $\alpha + \beta$ region may be referred to as a solid state miscibility gap that extends across the system. In some systems, a two-phase region of $\alpha + \beta$ may not span the system completely, but may close within the system in a manner essentially analogous with the case of the liquid miscibility gap. Two examples are illustrated in Figures 4.26–4.29; the first example is a ternary system in which binary system AB is eutectic in nature, showing partial solid solubility while binary systems BC and AC show complete solid solubility; in the other example, system AB is assumed to be peritectic in nature. Both of these ternary systems contain a solid state $\alpha + \beta$ region which extends only part of the way across the system, and isothermal sections through the region at, and below the temperature of point N, will show a critical point. The three-phase space $(L + \alpha + \beta)$ which rests on the top of the $\alpha + \beta$ region, also closes within the system.

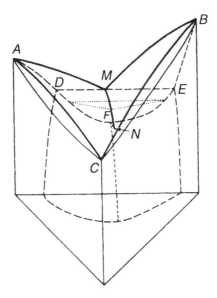

Fig. 4.26 Space model of a system containing a eutectic reaction in which the solid state miscibility gap closes within the system.

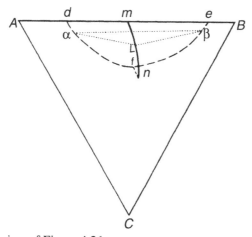

Fig. 4.27 Projected view of Figure 4.26.

The liquidus curves giving the compositions of liquid involved in either the eutectic or the peritectic reactions are shown in relation to the tie-triangles, representing the co-existence of L, α, and β. The changes of α and β compositions as the three-phase reaction proceeds, occur in a manner analogous to the changes of the liquid compositions

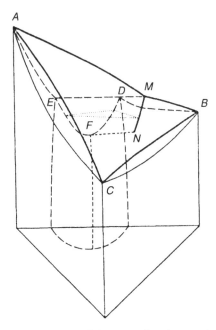

Fig. 4.28 Space model of a system containing a peritectic reaction in which the solid state miscibility gap closes within the system.

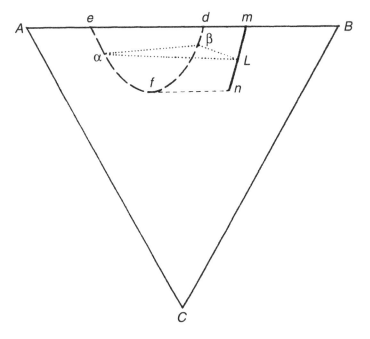

Fig. 4.29 Projected view of Figure 4.28.

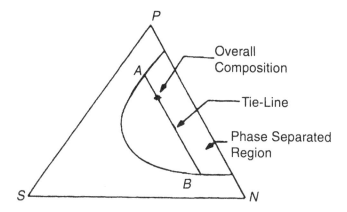

Fig. 4.30 Schematic isothermal section for a polymer-solvent-nonsolvent (P-S-N) system; the ends of the tie line within the miscibility gap shows the compositions of a concentrated polymer phase, *A*, and a dilute polymer phase, (*B. L. H. Sperling, Introduction to Physical Polymer Science*, second edition. Copyright 1992, John Wiley and Sons Inc. Reprinted by permission of John Wiley and Sons Inc.).

in a monotectic reaction, and the positions of tie-triangles relative to alloy composition points form a basis for discussing the progress and completion of the reaction. For example, in Figure 4.26 during the eutectic reaction the liquid composition changes along curve *MN* while the compositions of the solid solutions change along *DF* and *EF* respectively. There are some special features of each of these systems regarding the shape of the three-phase space, and the solidification of certain alloys which, although single phase when solid, undergo the three-phase reaction during solidification. Reference should be made to other accounts for details.[2,4,6]

4.7 MISCIBILITY GAPS IN TERNARY POLYMERIC SYSTEMS

In this section, as a sequel to the consideration in Chapter 2 of binary miscibility gaps in binary polymer systems, examples of miscibility in ternary polymeric systems are presented in the form of schematic isothermal sections at constant pressure.[17]

Figure 4.30 illustrates liquid–liquid phase separation in a system consisting of a polymer in a solvent–nonsolvent mixture. The polymer-solvent binary system shows complete intersolubility, while the polymer–nonsolvent solution shows a region of phase separation which extends into the ternary system. For compositions within this region, as shown, a tie-line joins the composition of a polymer–rich phase (*A*) and a dilute polymer phase (*B*).

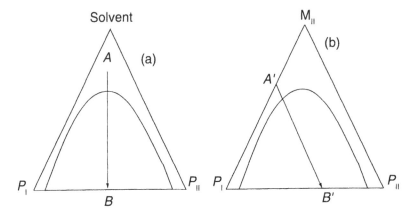

Fig. 4.31 Schematic isothermal systems. (a) System containing two types of polymer, which form a miscibility gap, and a solvent; the line *AB* illustrates that progressive decrease in solvent content leads to phase separation, (b) System containing two types of polymer, which form a miscibility gap, and a monomer; the line *A'B'* illustrates how the ternary miscibility gap, is entered and phase separation occurs as polymerisation of M_{II} to form P_{II} proceeds, (L. H. Sperling, *Introduction to Physical Polymer Science*, second edition Copyright 1992. John Wiley and Sons Inc. Reprinted by permission of John Wiley and Sons Inc.).

Figure 4.31a shows a schematic isothermal section of a system involving two polymers, P_I and P_{II}, and a solvent. On moving from point *A* to *B*, i.e corresponding to a decrease in solvent content, when the miscibility gap is entered, phase separation begins and proceeds to point *B*, where no solvent is present.

Another interesting and industrially important case is illustrated in Figure 4.31b, consisting of two polymers, P_I and P_{II}, and a monomer M_{II}. The respective binary systems between the polymers and the monomer show complete liquid miscibility, while the binary system consisting of the two polymers contains a miscibility gap. In a binary polymer-monomer blend, of composition *A'*, when the monomer (M_{II}) undergoes polymerisation to form polymer P_{II}, a change in composition occurs as represented along the line *A'B'*; phase separation occurs when the two phase region is entered and proceeds up to composition *B'*. Examples of this type of situation are materials such as high-impact polystyrene and *ABS* resins.

4.8 PROBLEMS

1. With reference to Figure 4.1:
a. State what phase regions are separated by the following surfaces:
     i. *ADGC*,

Table 4.1

Temperature	α	β	L
540°C	85% A, 10% B, 5% C	5% A, 93% B, 2% C	55% A, 30% B, 15% C
510°C	82% A, 11% B, 7% C	6% A, 89% B, 5% C	48% A, 32% B, 20% C

Table 4.2

	Temperature of Invariant Reaction	Compositions of the Phases Involved in Invariant Reaction		
		α	β	Liquid
System AB	850°C	95% A, 5% B	10% A, 90% B	50% A, 50% B
System AC	750°C	97% A, 3% C	12% A, 88% C	60% A, 40% C

    ii. *BMN*,

    iii. *DMNG*,

    iv. *DEFG*.

b. State what surfaces separate the following phase regions:

    i. $L + \beta/\beta$,

    ii. $L + \beta/\alpha + \beta + L$.

2. In the system *ABC*, a ternary alloy containing 40% *B* and 8% *C* undergoes a eutectic reaction $L \rightarrow \alpha + \beta$ over the temperature range 550–500°C as part of its solidification sequence. The compositions of the phases coexisting in equilibrium at 540 and 510°C are shown in Table 4.1.

Calculate:

a. The proportions by weight of α and β phases present in the alloy at 540°C.

b. The ratio of the proportions of liquid phase present at 540 and 510°C.

3. (a) With reference to Figure 4.8 describe the solidification sequence of alloy *X*; refer also to Figure 4.1 to describe the compositional changes that occur during solidification.

b. On what surfaces respectively of Figure 4.1 do the points T_c and T_s shown in Figure 4.8 lie?

4. A ternary system *ABC* shows complete solubility in the liquid state and contains only two solid phases, which are solid solutions designated α and β respectively. The melting points of components *A*, *B* and *C* are 1100, 900 and 800°C, respectively. The binary system *BC* shows complete solid solubility, while systems *AB* and *AC* each contain an invariant reaction given in Table 4.2.

Table 4.3

Temperature	Phase Compositions		
	Liquid	α	β
550°C	69% A, 19% B, 12% C	57% A, 41% B, 2% C	20% A, 78% B, 2% C
520°C	66% A, 18% B, 16% C	56% A, 40% B, 4% C	19% A, 78% B, 3% C
500°C	63% A, 17% B, 20% C	55% A, 39% B, 6% C	18% A, 78% B, 4% C

Sketch and label:

a. Possible liquidus and solidus projections for the ternary system.

b. By reference to these projections, describe the solidification sequence of an alloy containing 30% A, 55% B and 15% C, under equilibrium conditions; estimate the proportions of the phases present at the completion of solidification.

5. With reference to Figure 4.9:

a. State what phase regions are enclosed by the following surfaces:

i. *AME, AEF, AMN, ANF* and *EMNF*;

ii. *MBD, MBCN, NCG* and *DBCG*;

iii. *EDGF, DMNG* and *EMNF*.

b. State what surfaces separate the following phase regions:

i. $\alpha + \beta/L + \alpha + \beta$,

ii. $L/L + \alpha$.

6. A ternary system *ABC* shows complete liquid solubility and contains only two solid phases, namely two solid solutions designated α and β respectively.

Table 4.3 shows data refering to the compositions of liquid, α and β coexisting in equilibrium at the temperatures.

a. Calculate the equilibrium percentages of liquid, α and β respectively, present in an alloy containing 44% A, 50% B, and 6% C at 520°C.

b. Deduce, giving your reasoning, the nature of the three phase reaction undergone by the alloy containing 44% A, 50% B and 6% C during solidification.

c. Sketch a projected view of the liquidus, and an isothermal section in the solid state for a system consistent with the data given above: also sketch an isothermal section for 550°C.

d. With reference to the sketches of (c) describe fully the solidification sequence you would expect for an alloy containing 60% A, 30% B and 10% C, assuming equilibrium conditions.

7. Making use of data obtained from Figure 4.22c, calculate how the proportion of Pb-rich liquid will vary with Sn content in the range 0–25 wt.% Sn in a series of Pb–Zn–Sn alloys each containing 55 wt.% Pb, equilibrated at 520°C.

5. Systems Containing Four Phases

5.1 INTRODUCTION

Invariant reactions in ternary systems are associated with the occurrence of four phases co-existing in equilibrium. Rhines[6] has classified four-phase equilibria into three types or classes. This classification is followed here and can be illustrated by considering the three classes in terms of a liquid phase (L) and three solid solution phases(α, β and γ):

Class I: $L \rightleftharpoons \alpha + \beta + \gamma$

i.e. a ternary eutectic reaction.

Class II: $L + \alpha \rightleftharpoons \beta + \gamma$

This type of reaction can be thought of as being intermediate between eutectic and peritectic types; it is sometimes (as in the present work) referred to as a peritectic type, although Rhines does not favour this terminology.[*]

Class III: $L + \alpha + \beta \rightleftharpoons \gamma$

This class of reaction is designated as ternary peritectic in nature.

Considering here the situations of invariant ternary equilibria and cooling reactions involving a liquid phase and three solid phases, each of the three classes of reaction is associated with the intersection of three liquidus curves; each of these curves represents the path along which the liquid composition changes during a three phase-reaction during solidification (e.g. $L \rightarrow \alpha + \beta$).

The three classes can be defined in terms of the number of three-phase equilibria above and below the invariant temperature. As discussed in detail by Rhines, there are many possible combinations of type of phase giving rise to the three classes of invariant equilibria and reactions: for example,involving a liquid phase (or phases) with solid phases; solid phases only; combinations including a gas phase.

In the context of solidification reactions, the principles of using ternary liquidus projections are discussed further in Chapter 7, taking account of the intersecting liquidus curves and the number of three phase equilibria above and below the invariant temperature.

---

[*] Although the class II reaction is referred to here as a peritectic it should be noted that in modern assessments of phase diagrams this type of reaction is normally designated by the letter U (German 'Ubergang') to distinguish it from the Class III peritectic (designated P).

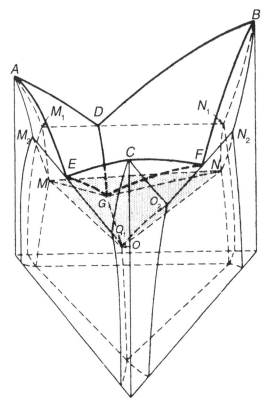

Fig. 5.1 Space model of system showing a ternary eutectic reaction $L \rightarrow \alpha + \beta + \gamma$ (Class I).[6]

5.2 SYSTEMS CONTAINING A TERNARY EUTECTIC REACTION (CLASS I)

5.2.1 A SYSTEM IN WHICH THE THREE BINARY SYSTEMS ARE OF THE EUTECTIC TYPE

Figure 5.1 shows the space model of a ternary system ABC in which each of the constituent binary systems is of the eutectic type, with partial solid solubility of the components: the primary solid solutions are α, β, and γ respectively and no intermediate phases are present. Eutectic valleys (DG, FG, and EG) run from the binary systems to lower temperatures to meet at the ternary eutectic point G. For each of these eutectic reactions there exists a three-phase region made up of tie-triangles. For the $L \rightarrow \alpha + \beta$ reaction, curve DG represents the composition of the liquid, curve M_1M the composition of α, and N_1N the composition of β; for the $L \rightarrow \beta + \gamma$ reaction, the liquid composition lies on FG, and the β and γ compositions on N_2N and O_2O respectively, while for the $L \rightarrow \alpha + \gamma$ reaction, the liquid composition lies on EG and the α and γ compositions on M_2M and O_1O respectively.

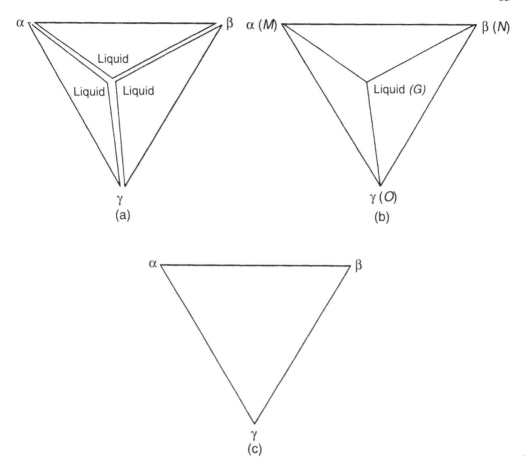

Fig. 5.2 Tie-triangles (a) at a temperature just above that of the ternary eutectic, (b) at the ternary eutectic temperature and (c) below the ternary eutectic temperature.

With decreasing temperature the positions of the tie-triangles change as the compositional changes occur in the phases involved in the reaction. The ternary eutectic reaction occurs at the temperature at which the liquid composition corners of these three triangles become coincident with one another at the ternary eutectic point (G in Figure 5.1); at the invariant reaction temperature the $L\alpha$ sides of the triangles coincide, as do also the $L\beta$ and $L\gamma$ sides respectively. Figure 5.2a indicates the relative positions of the tie-triangles at a temperature just above the ternary eutectic, and Figure 5.2b shows the merging of the triangles at the ternary eutectic temperature. The ternary eutectic reaction is represented by a plane in the space model (MNO in Figure 5.1); the composition of each of the four phases participating in the reaction is fixed, and is therefore represented by a point on the plane, and the liquid composition point lies within the triangle formed

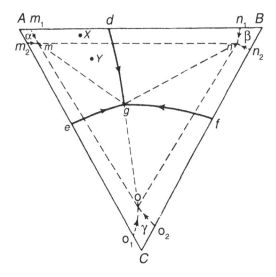

Fig. 5.3 Projected view of system of Figure 5.1.

by the compositions of the three solid phases. Under equilibrium conditions, the temperature remains constant during the eutectic reaction until all the liquid is consumed.

Below the ternary eutectic plane there lies a three-phase region, $\alpha + \beta + \gamma$, and the equilibrium state at a temperature within this region is represented by the $\alpha\beta\gamma$ tie-triangle shown in Figure 5.2c. During cooling in the solid state, solubility changes generally occur, and the composition co-ordinates of the α, β, and γ corners will therefore change; however, it is convenient to neglect these changes, although they are indicated in Figure 5.1.

Thus, the ternary eutectic reaction involves three three-phase regions that approach the eutectic plane from higher temperatures, and one three-phase region existing below the plane.

5.2.1.1 Projected View of the System

Figure 5.3 shows a projected view of the system. Curves *dg*, *fg*, and *eg* are the liquidus curves, the eutectic nature of the invariant reaction being indicated by the arrows on the curves all pointing towards *g*. Regions *Adge*, *Bdgf*, and *Cegf* define the primary phase solidification, i.e. of α, β, and γ respectively. The solidus curves m_1m and m_2m define the boundaries of the α-solid solution region at the completion of solidification, while the β and γ single-phase regions are bounded by n_1n and n_2n, and o_1o and o_2o respectively. The triangle *mno* is the projection of the isothermal eutectic plane and on its upper surface is divided into the three tie-triangles *mng*, *mgo*, and *ngo*; the tie-triangle, *mno* defines the three-phase region in the solid state while the solid state constitution is completed by the two-phase regions of $\alpha + \beta$, $\beta + \gamma$, and $\alpha + \gamma$ respectively (m_1n_1nm, nn_2o_2o, and m_2moo_1).

If solid state solubility changes are neglected, Figure 5.3 serves to define the limits of the solid state regions.

5.2.1.2 The Solidification of Alloys

The solidification of alloys in the various regions of the system may be discussed on the basis of Figure 5.3 assuming equilibrium conditions. If an alloy lies in one of the single-phase regions, solidification occurs in a manner typical of a solid-solution type alloy as described in a previous section. For alloys lying in one of the two-phase regions but not on a eutectic valley, the primary phase solidification occurs, either α, β, or γ, depending on the primary field in which the alloy composition lies. As a result of the separation of the primary phase, the liquid composition reaches one of the eutectic valleys, and the appropriate reaction begins; solidification is completed with falling temperature, leaving a mixture of the two solid phases. In alloy X, for example, the primary separation of α is followed by the reaction $L \rightarrow \alpha + \beta$: during the reaction the liquid corner of each tie-triangle lies on curve dg, and the α and β corners on curves m_1m and n_1n respectively, and solidification is complete when the $\alpha\beta$ side of the tie-triangle contains point X.

If the alloy composition lies in the three-phase region the liquid composition reaches the ternary eutectic point g, e.g. in alloy Y, as a result of the occurrence of the $L \rightarrow \alpha + \beta$ eutectic; the $L\alpha\beta$ tie-triangle has moved down on to the ternary eutectic plane without the alloy composition point coming to lie on the $\alpha\beta$ side of the triangle. At the ternary invariant temperature, the $L\alpha\beta$ tie-triangle is mng, and the reaction $L \rightarrow \alpha + \beta + \gamma$ occurs, with the liquid composition represented by point g, and the α, β, and γ compositions by m, n, and o respectively.

In the type of system discussed, which shows partial solid solubility, tracing of the actual path of the liquid composition on the liquidus surface during the primary stage of solidification requires tie-line data. This aspect has been previously discussed.

However, if it is assumed that the system shows no solid solubility, the compositional changes, which occur during solidification, may be accurately traced. Figure 5.4 represents the liquidus projection of such a system; all ternary alloys consist of $A + B + C$ in the solid state. The regions of the primary solidification of A, B, and C respectively are $Adge$, $Bdgf$, and $Cegf$ respectively.

The extremities of the tie-lines representing the liquid compositions at successively lower temperatures form the path of the liquid composition on the liquidus surface (Figure 5.4). Since the composition of the primary solid at all temperatures is the same, the liquid composition path is, in fact, the extension of the straight line joining the composition of the solid and the alloy composition (e.g. AX in Figure 5.4). When the liquid composition reaches a point on one of the binary eutectic valleys the primary solidification ceases (e.g. for alloy X, this occurs at point Z on curve dg). (At this temperature, prior to the commencement of the eutectic reaction, the amounts of liquid and solid co-existing can be calculated from the tie-line AXZ; thus the amount of primary A is given by the ratio XZ/AZ). Then, with falling temperature the eutectic reaction commences and, as the two solids separate, the liquid composition moves down the

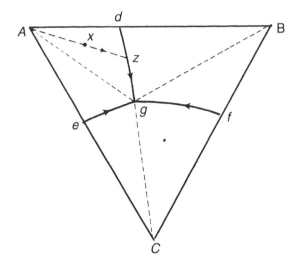

Fig. 5.4 Projected view of ternary eutectic system $L \rightarrow A + B + C$.

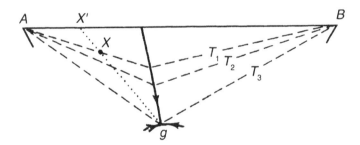

Fig. 5.5 Tie-triangles illustrating the progress of the eutectic reaction $L \rightarrow A + B$ in the system of Figure 5.4.

eutectic valley (e.g. for alloy X, along portion Zg of curve dg, while A and B separate). The progress of the eutectic reaction can be represented by tie-triangles at decreasing temperatures (T_1, T_2, and T_3) in Figure 5.5; since the compositions of the solids do not change during the reaction, the side of the triangle joining the two solids (i.e. AB in Figure 5.5) cannot move so as to contain the alloy composition point X. Solidification cannot be completed until the liquid composition reaches the ternary eutectic point g, when the reaction $L \rightarrow A + B + C$ occurs, (The percentage of liquid that finally solidifies as the ternary eutectic can be calculated from the tie-triangle AgB (Figure 5.5): this percentage is given by $\dfrac{XX'}{gX'} \times 100$)

Referring to Figure 5.4, if the alloy composition lies on one of the three straight lines joining the respective components to the ternary eutectic point (e.g. line Ag), the primary solidification stage is followed directly by the ternary eutectic reaction. Alloys in the region Adg undergo the eutectic reaction $L \rightarrow A + B$ during solidification, while alloys in region Aeg undergo the reaction $L \rightarrow A + C$: similarly, lines Bg and Cg divide the other primary fields in a corresponding way.

In systems where there is solid solubility, but very limited in extent, the liquid composition change may be estimated by the method described above for the case of no solid solubility, i.e. by drawing the extension of the straight line from the appropriate corner of the system through the alloy point. However, this procedure is only an approximation where solid solubility exists, and should not be used where the solid solubility is extensive.

5.2.1.3 Isothermal Sections

Figures 5.6a and 5.6b, show two isothermal sections through the system with partial solid solubility (Figure 5.1). The section shown in Figure 5.6a is at a temperature just above the ternary eutectic temperature, but below the temperature of the eutectic reactions in the three binary systems. The three triangles representing the $L\alpha\beta$, $L\beta\gamma$, and $L\alpha\gamma$ regions are seen in relation to the adjoining single-phase and two-phase regions, with decreasing temperature, the triangles move down on to the ternary eutectic plane as described earlier. Figure 5.6b shows a section at a temperature just below that of the eutectic reaction: if solid state solubility changes are neglected, the limits of the solid phase regions in this isothermal section will coincide with those shown in the projected view (Figure 5.3).

In the solid state isothermal section (Figure 5.6b) thermodynamic considerations lead to the following principle regarding the boundaries between the single-phase and two-phase regions where they meet the three-phase region; if these boundaries are extrapolated, then they must either both lie within the three-phase region, or must project into two different two-phase fields, as shown for example in Figures 5.7a and b.[6, 22] Incorrect constructions are shown in Figures 5.7c and d; in the former case one of a pair of boundaries projects into the three-phase region and the other into a two-phase region, while in the latter both boundaries are shown as projecting into the single-phase region.

5.2.2 A SYSTEM IN WHICH ONE OF THE BINARY SYSTEMS IS OF THE PERITECTIC TYPE AND TWO ARE OF THE EUTECTIC TYPE

If one of the binary systems is of the peritectic type, the ternary invariant reaction can still be of the eutectic type; the binary peritectic may change to eutectic as it moves into the system. An example of this is discussed later with reference to a system containing an intermediate phase (see page 91).

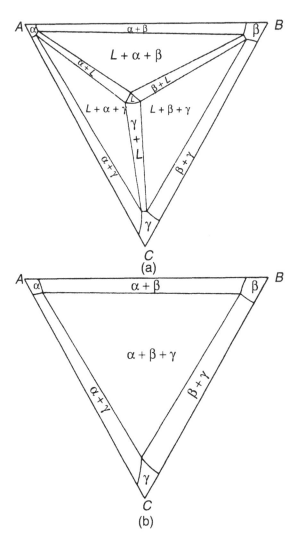

Fig. 5.6 (a) Isothermal sections through the system of Figure 5.1 at a temperature above the ternary eutectic temperature, but below the temperatures of the eutectic in the three binary systems and (b) at a temperature below that of the ternary eutectic.

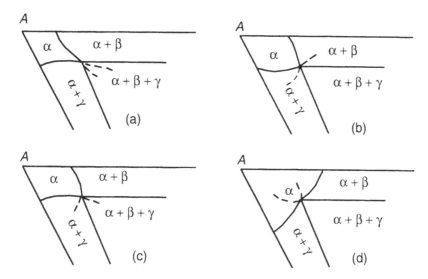

Fig. 5.7 (a) and (b) correct and (c) and (d) incorrect constructions for boundaries of single-phase regions.

5.3 SYSTEMS CONTAINING A TERNARY PERITECTIC REACTION (CLASS II AND CLASS III)

5.3.1 A System in Which One Binary System is of the Peritectic Type, and Two are of the Eutectic Type, and Which Shows an Invariant Reaction of the Type $L + \alpha \rightarrow \beta + \gamma$ (Class II)

In the ternary system ABC illustrated in Figure 5.8, the system AB is peritectic in nature, and systems BC and AC are eutectic in nature. Terminal solid solutions α, β and γ and are formed, and there are no intermediate phases. (The type of invariant reaction depicted can also occur with other combinations of binary systems).[4]

The curve DG running into the ternary system represents the path of the composition of liquid taking part in the $L + \alpha \rightarrow \beta$ peritectic reaction. Curve EG is the liquid composition curve for the eutectic reaction $L \rightarrow \alpha + \gamma$. Point G, where these curves intersect, represents the composition of liquid taking part in the invariant peritectic reaction $L + \alpha \rightarrow \beta + \gamma$. From G, the eutectic valley, GF, for the reaction $L \rightarrow \beta + \gamma$ runs to lower temperatures.

Associated with each of the curves there exists a three-phase region made up of tie-triangles. For the $L\alpha\beta$ region, curve M_1M represents the α-phase composition, and N_1N the composition of β; for the $L\alpha\gamma$ region the α and γ compositions lie along curve M_2M and O_1O respectively: for the $L\beta\gamma$ region the β and γ compositions lie along NN_2 and OO_2 respectively.

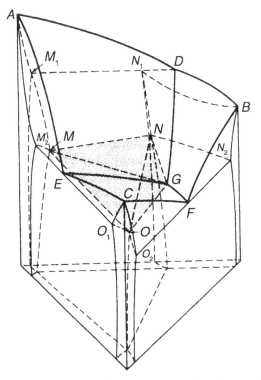

Fig. 5.8 Space model of system showing a ternary peritectic reaction $L + \alpha \rightarrow \beta + \gamma$ (Class II).[6]

The ternary peritectic reaction may be considered on the basis of the tie-triangles making up the three-phase regions. At a temperature just above that of the ternary peritectic reaction, the relative positions of the $L\alpha\beta$ and $L\alpha\gamma$ tie-triangles are shown in Figure 5.9a. With decreasing temperature the positions of the triangles change as the compositional changes occur in the phases, and, in effect, the triangles are displaced towards one another. At the ternary peritectic temperature, the liquid composition reaches point G, and the α composition reaches point M, so that the $L\alpha$ sides of the triangles become coincident (Figure 5.9b), and the two triangles form a trapezium.

At the peritectic temperature four phases L, α, β, and γ co-exist, the compositions of these phases lying at the corners of the trapezium $MNGO$ (see Figure 5.8). This trapezium may be termed the peritectic plane. It should be noted that the liquid composition point, G, lies outside the triangle formed by joining the points representing the compositions of the solid phases α, β, and γ (i.e. points M, N, and O); this arrangement of the phase composition points distinguishes the peritectic reaction from the invariant eutectic reaction. The peritectic reaction occurs by the interaction of L and α (whose compositions lie at opposite corners of the trapezium) to form a mixture of the two phases (β and γ) whose compositions lie at the other opposite corners of the trapezium.

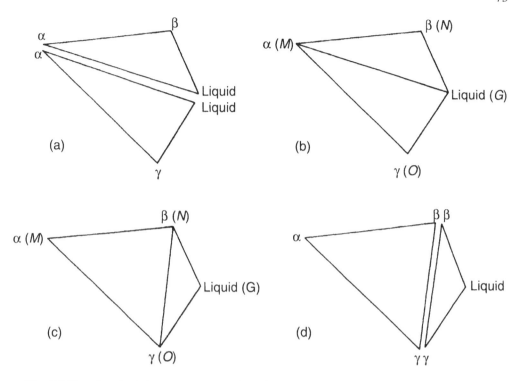

Fig. 5.9 Tie-triangles (a) at a temperature just above that of the ternary peritectic, (b) and (c) at the ternary peritectic temperature and (d) below the ternary peritectic temperature.

At the completion of the invariant reaction, the trapezium is divided across the βγ diagonal to form two triangles which represent the phase equilibria between αβγ and Lβγ respectively (Figure 5.9c). Thus, there exist two three-phase regions above the peritectic plane (namely Lαβ and Lαγ) and two below the plane (namely αβγ and Lβγ).

With falling temperature, the two tie-triangles αβγ and Lβγ separate from one another, as compositional changes occur in the phases (Figure 5.9d). The Lβγ triangle is displaced as the liquid composition changes along the eutectic curve GF, while the β and γ compositions change along curves NN_2 and OO_2 respectively.

Alloys whose compositions lie within the αβγ triangle (MNO) at the peritectic temperature, complete their equilibrium solidification process by the ternary peritectic reaction, since all the liquid is consumed in the reaction. However, for alloys whose compositions lie in the Lβγ triangle (GNO) at the peritectic temperature, the α-phase is consumed in the invariant reaction; then with further decrease in temperature, solidification is finally completed by the displacement of the Lβγ triangle until the βγ side of the triangle contains the alloy composition point. An alloy whose composition corresponds to the intersection of the Lα and βγ diagonals of the trapezium at the peritectic temperature is composed entirely of L and α just above this temperature, and entirely of β and γ just below it.

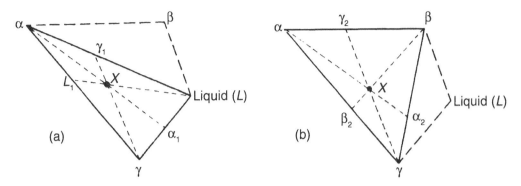

Fig. 5.10 Tie-triangles at the ternary peritectic temperature used to calculate the percentages of the phases present in alloy X before and after the occurrence of the invariant reaction.

To illustrate the nature of the invariant reaction it is helpful to calculate the percentages of the phases present in a typical alloy just above and below the peritectic temperature. For example, in alloy X (Figure 5.10a) at the peritectic temperature at the beginning of the invariant reaction:

$$\%\text{Liquid} = \frac{L_1 X}{LL_1} \times 100, \quad \%\alpha = \frac{\alpha_1 X}{\alpha\alpha_1} \times 100, \quad \%\gamma = \frac{\gamma_1 X}{\gamma\gamma_1} \times 100,$$

no β is present.

After the invariant reaction is completed (Figure 5.10b):

$$\%\alpha = \frac{\alpha_2 X}{\alpha\alpha_2} \times 100$$

(i.e. less than at the beginning of the reaction),

$$\%\gamma = \frac{\gamma_2 X}{\gamma\gamma_2} \times 100$$

(i.e. a greater amount than at the beginning of the reaction),

$$\%\beta = \frac{\beta_2 X}{\beta\beta_2} \times 100,$$

no liquid is present.

Thus, during the reaction, the percentages of β and γ have increased at the expense of the L and α phases, in accordance with the peritectic nature of the process $L + \alpha \rightarrow \beta + \gamma$.

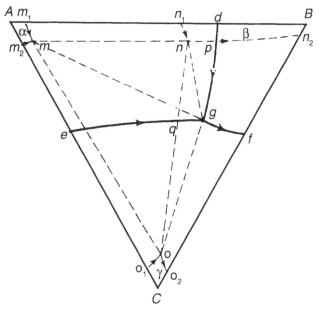

Fig. 5.11 Projected view of system of Figure 5.8.

5.3.1.1 Projected View of the System

Figure 5.11 shows a projected view of the system. The liquidus curves are *dg, eg,* and *gf,* and the peritectic nature of the invariant is reaction is indicated by the arrows on two of the curves pointing towards the intersection point of the curves, while the arrow on the third curve points away from the intersection. The primary separation of α, β, and γ occurs within the regions *Adge, Bdgf,* and *Cegf* respectively. The solidus curves, m_1m and m_2m define the limits of the α-solid solution region, while the β and γ solid solution regions are bounded by curves n_1n, nn_2 and o_1o, oo_2 respectively. The trapezium *mngo* is the projection of the isothermal plane representing the invariant reaction. As shown in Figures 5.9b and c the upper surface of the plane is made up of the tie-triangles *mng* and *mog* while the lower surface is made up of *mno* and *ngo.* All alloys whose compositions lie within the trapezium undergo the peritectic reaction.

In the solid state part of the system, the triangle *mno* (which is part of the solidus of the system) defines the three-phase α + β + γ region at a temperature just below the peritectic temperature. The two-phase regions α + β, α + γ, and β + γ are $m_1n_1nm, m_2moo_1,$ and nn_2o_2o respectively. The same types of solid state region are in fact present as in the ternary eutectic system previously discussed, but the position of the invariant liquid composition, *g,* relative to the positions of the α, β and γ phases (*m, n,* and *o* respectively) is different; i.e. point *g* lies outside the triangle *mno,* not inside it as in a ternary eutectic.

5.3.1.2 The Solidification of Alloys

For most of the alloys lying in one of the single phase regions of Figure 5.11 no detailed consideration of the solidification sequence is necessary here. However, it should be noted that in the region n_1dpn, solidification commences with the formation of primary α, and there follows the peritectic reaction $L + \alpha \rightarrow \beta$ in which all the α is consumed; the liquid composition, then changes along a path on the β-liquidus surface $Bdgf$, and β is deposited until solidification is completed.

With regard to the two-phase regions, alloys within the region m_1n_1nm commence their solidification by the formation of primary α, then follows the reaction $L + \alpha \rightarrow \beta$ when the liquid composition has reached a point on curve dg; solidification is completed when the $\alpha\beta$ side of the $L\,\alpha\beta$ tie-triangle contains the alloy composition point. Alloys in the $\alpha + \gamma$ region (i.e. m_2moo_1) begin to solidify by depositing either α or γ, depending on the actual alloy composition; primary solidification is followed by the binary eutectic reaction $L \rightarrow \alpha + \gamma$ that proceeds until all the liquid is consumed.

The $\beta + \gamma$ region (i.e. nn_2o_2o) is best considered in several sections:

pn_2fg: Solidification begins by the formation of primary β (i.e. Liquid $(L) \rightarrow \beta$) and then the liquid composition changes along a path on the β-liquidus until it reaches the eutectic curve gf. The reaction $L \rightarrow \beta + \gamma$ occurs as the liquid composition changes along gf, and the β and γ compositions change along nn_2 and oo_2 respectively, until the liquid is consumed.

npg: The first stage of solidification is $L \rightarrow \alpha$, and the liquid composition then changes until it reaches curve dg when the reaction $L + \alpha \rightarrow \beta$ occurs. Before the liquid composition reaches point g, the α-phase will have been consumed, and the liquid composition then leaves curve dg and changes along a path on the β-liquidus while β is deposited. The liquid composition eventually reaches the eutectic curve gf when the solidification proceeds as for region pn_2fg, by the deposition of β and γ.

oo_2fg: The primary stage of solidification is $L \rightarrow \gamma$, and the liquid composition changes on to curve gf when the $L \rightarrow \beta + \gamma$ reaction occurs as for the region pn_2fg.

ngo: Alloys in this region undergo the invariant peritectic reaction during solidification. The primary stage of solidification is, either $L \rightarrow \alpha$ (in region ngq) or $L \rightarrow \gamma$ (in region oqg) and the liquid composition then changes either on to the eutectic curve eg (i.e. $L \rightarrow \alpha + \gamma$) or on to the peritectic curve dg (i.e. $L + \alpha \rightarrow \beta$) depending on the alloy composition; with falling temperature the appropriate three-phase reaction proceeds until the liquid composition reaches the invariant peritectic point g when the reaction $L + \alpha \rightarrow \beta + \gamma$ occurs. The α-phase is consumed, leaving liquid, β and γ, and with further decrease in temperature the $L \rightarrow \beta + \gamma$ reaction proceeds along curve gf, the final structure being $\beta + \gamma$.

The remaining region to be considered is the three-phase region mno. For alloys within this region the primary separation may be either of α or of γ, and the liquid composition moves either on to curve eg or curve dg, depending on the actual alloy composition. As the three-phase reaction (i.e, either $L \rightarrow \alpha + \gamma$ or $L + \alpha \rightarrow \beta$) proceeds the liquid composition reaches point g, when the invariant reaction occurs and all the liquid is consumed under equilibrium conditions.

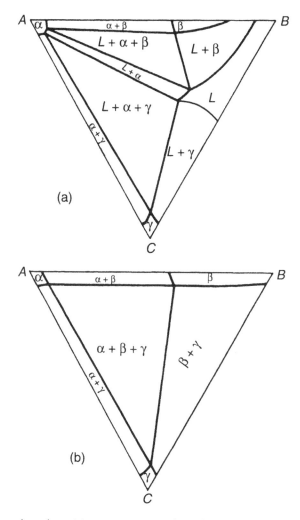

Fig. 5.12 Isothermal sections (a) at a temperature just above that of the ternary peritectic (b) at a temperature below that of the eutectic reaction in system *BC*.

5.3.1.3 Isothermal Sections

Figure 5.12a shows an isothermal section at a temperature just above the peritectic temperature. The three-phase triangles $L + \alpha + \beta$ and $L + \alpha + \gamma$, associated with the three-phase reactions $L + \alpha \rightarrow \beta$ and $L \rightarrow \alpha + \gamma$ are seen in relation to the adjacent regions.

The temperature of the isothermal section shown in Figure 5.12b is below that of the eutectic reaction in system *BC*, and is representative of the solid state constitution of the system. Assuming that no solid state solubility changes occur after solidification is

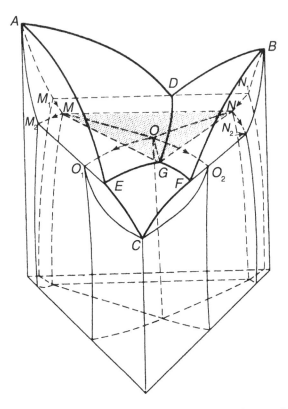

Fig. 5.13 Space model of system showing a ternary peritectic reaction $L + \alpha + \beta \rightarrow \gamma$ (Class III).[6]

complete, the limits of the phase regions in this section coincide with those shown in the projected view of the system (Figure 5.11).

5.3.2 A SYSTEM IN WHICH ONE BINARY SYSTEM IS OF THE EUTECTIC TYPE, AND TWO ARE OF THE PERITECTIC TYPE, AND WHICH SHOWS AN INVARIANT REACTION OF THE TYPE $L + \alpha + \beta \rightarrow \gamma$ (CLASS III)

In the ternary system ABC illustrated in Figure 5.13, system AB is of the eutectic type and systems BC and AC are peritectic in nature. Terminal solid solutions α, β and γ are formed and there are no intermediate phases. (The type of invariant reaction depicted can also occur with other combinations of binary systems).[4]

The eutectic valley DG $(L \rightarrow \alpha + \beta)$ runs down into the ternary system. Point G represents the composition of liquid taking part in the invariant peritectic reaction $L + \alpha + \beta \rightarrow \gamma$. From G, running to lower temperatures, are two peritectic curves GE $(L + \alpha \rightarrow \gamma)$ and GF $(L + \beta \rightarrow \gamma)$.

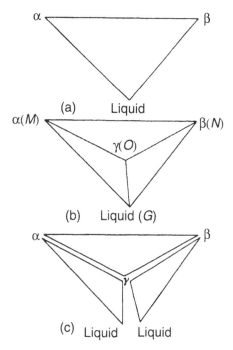

Fig. 5.14 Tie-triangles (a) at a temperature just above that of the ternary peritectic (b) at the peritectic temperature and (c) below the peritectic temperature.

For the eutectic reaction $L \rightarrow \alpha + \beta$, as the liquid composition changes along DG, the α and β compositions change along M_1M and N_1N respectively. For the peritectic reaction $L + \alpha \rightarrow \gamma$, the compositions of the liquid, α and γ change along the curves GE, MM_2, and OO_1 respectively, while for the $L + \beta \rightarrow \gamma$ reaction, the compositional changes of liquid, β and γ occur along GF, NN_2, and OO_2 respectively.

The essential nature of the invariant reaction may be considered on the basis of the tie-triangles that constitute the three-phase regions in the system. One three-phase region namely $L + \alpha + \beta$, extends down on to the plane that represents the invariant reaction. A typical $L\alpha\beta$ tie-triangle through this space, at a temperature just above the peritectic, is shown in Figure 5.14a. At the invariant reaction temperature, this triangle (GMN) defines the compositions of the liquid, α and β that take part in the reaction. The composition of the γ-phase lies at point O, within the triangle GMN; thus the liquid composition G lies outside the triangle (MNO) formed by the compositions of the three solids, which is a characteristic feature of the peritectic reaction. The reaction gives rise to three three-phase regions below the peritectic plane, represented by the tie-triangles $\alpha\beta\gamma$ (MNO), L $\alpha\gamma$ (GMO), and $L\beta\gamma$ (GNO) in Figure 5.14b. With decreasing temperature these three tie-triangles become separated as compositional changes occur (Figure 5.14c); the triangles $L\alpha\gamma$ and $L\beta\gamma$ are associated with the reaction $L + \alpha \rightarrow \gamma$ and $L + \beta \rightarrow \gamma$ respectively.

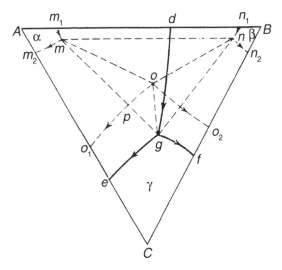

Fig. 5.15 Projected view of system of Figure 5.13.

Alloys whose compositions lie within the $\alpha\beta\gamma$ region at the peritectic temperature complete their solidification by the invariant reaction, all the liquid being consumed. However, if the alloy composition lies either in the $L\,\alpha\,\gamma$ or $L\,\beta\gamma$ region of the peritectic plane, then some liquid remains at the completion of the invariant reaction, and the solidification sequence continues via either the $L + \alpha \rightarrow \gamma$ or the $L + \beta \rightarrow \gamma$ reaction.

5.3.2.1 Projected View of the System

Figure 5.15 shows a projected view of the system. The liquidus curves are dg, ge, and gf, and the type of peritectic reaction is indicated by the fact that the arrow on the first of these curves points to the intersection of the curves, and the arrows on the other two curves point away from the intersection. The primary separations of α, β, and γ occur within the regions $Adge$, $Bdgf$, and $Cegf$ respectively. The limits of the α, β, and γ regions are defined by the solidus curves m_1m, mm_2, n_1n, nn_2, and oo_2, oo_1 respectively. Triangle mng is the projection of the isothermal reaction plane, and all alloys whose compositions lie within this triangle undergo the peritectic reaction. The division of the triangle on the underside into the tie-triangles mno, mog, ngo is also shown. Triangle mno is part of the solidus of the system and in the solid state part of the system represents the three-phase $\alpha\,\beta\,\gamma$ region. The two-phase regions $\alpha + \beta$, $\alpha + \gamma$, and $\beta + \gamma$ are m_1n_1nm, m_2moo_1, and nn_2o_2o respectively.

5.3.2.2 The Solidification of Alloys

The solidification sequences for alloys that undergo the peritectic reaction (i.e. alloys within triangle *mng*) illustrate the main features of the system. Following the primary separation of α or β, the liquid composition moves on to the curve *dg*, and then along this curve as the reaction $L \rightarrow \alpha + \beta$ proceeds until the invariant point *g* is reached, when the reaction $L + \alpha + \beta \rightarrow \gamma$ occurs. If the alloy composition lies within triangle *mno* all the liquid is consumed in the reaction.

For alloys in the region *mog* all the β is consumed, leaving liquid, α and γ, and with falling temperature the peritectic reaction $L + \alpha \rightarrow \gamma$ proceeds, the liquid composition changing along *ge*, and the α and γ compositions along mm_2 and oo_1 respectively. If the alloy composition is within region *mop* so that the alloy consists of α + γ in the solid state, all the liquid is consumed in the reaction $L + \alpha \rightarrow \gamma$. For alloy compositions within region *ogp*, which are single phase γ when solid, all the α is consumed in the $L + \alpha \rightarrow \gamma$ reaction and the remaining liquid moves away from curve *ge*, over the γ liquidus surface, while γ is deposited until solidification is completed.

For alloys in the region *ngo*, the α is consumed in the invariant reaction, leaving liquid, β and γ, and solidification then continues via the $L + \beta \rightarrow \gamma$ reaction along curve *gf*, in a manner analogous to that for region *mog*.

The solidification of alloys in the remaining regions of the system is not discussed here; the general principles previously discussed for other systems are applicable.

5.3.2.3 Isothermal Section

Figure 5.16 shows an isothermal section in the solid state, and if solubility changes following solidification and neglected, the limits of the phase regions coincide with those on the projected view (Figure 5.15).

5.4 PROBLEMS

1. With reference to Figure 5.1:
a. State what phase regions are enclosed by the following surfaces:
i. BDN_1, $BDGF$, DN_1NG, FN_2NG, BN_1NN_2, BFN_2;
ii. M_2MGE, O_1OGE, M_2MOO_1, MGO.
b. State what surfaces separate the following phase regions:
i. $L + \alpha + \beta / L + \alpha$;
ii. $\gamma / L + \gamma$.
2. With reference to Figure 5.8:
a. State what surfaces enclose the following phase regions:
i. $L + \alpha + \beta$,

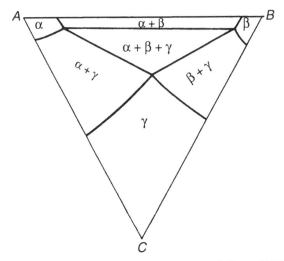

Fig. 5.16 Isothermal section in the solid state for system of Figure 5.13.

Table 5.1

Alloy Number	Alloy Composition			Temperatures (°C)
	wt.% A	wt.% B	wt.% C	
1	60	40	-	750
2	51	-	49	800
3	-	44	56	700
4	80	10	10	950 (Primary A); 600
5	30	50	20	850 (Primary B); 600

ii. $L + \alpha$.

b. State what phase regions are separated by the surfaces:

i. $OGFO_2$;

ii. MNO;

iii. NGO.

3. The melting points of metals, A, B and C are 1000, 950 and 850°C, respectively. The metals show complete liquid solubility, but negligible solid solubility in one another: they form no compounds.

Table 5.1 gives a complete list of the temperatures of cooling curve discontinuities and arrests for certain alloys.

a. Draw a liquidus projection for the ternary system ABC that conforms with the above data.

b. Calculate the proportion of primary A that forms during the solidification of the alloy containing 80% A, 10% B and 10% C.

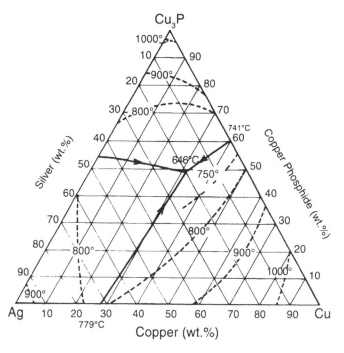

Fig. 5.17 Ag–Cu–Cu$_3$P system. (From G. H. Sistaire, Metals Handbook, 1973, Reproduced with permission from *ASM International*, Materials Park, OH 44073-0002, U.S.A.)

4. Ag–Cu–Cu$_3$P. Part of the Cu–Ag–P system is represented by the liquidus projection in Figure 5.17, taking the compound Cu$_3$P as one of the components. Each of the binary systems contains a eutectic and the ternary system contains an invariant eutectic $L \rightleftharpoons Ag + Cu + Cu_3P$. (Solid solubility, which is small at room temperature, is neglected here). The system includes a group of brazing alloys (in the range 0–15 wt.% Ag; 40-60 wt.% Cu; 35–55 wt.% Cu$_3$P) used for joining copper alloys.

a. For an alloy containing 50 wt.% Cu$_3$P and 5 wt.% Ag determine:
i. the liquidus temperature;
ii. the percentage of liquid present at the temperature when separation of Cu$_3$P begins;
iii. the proportions of the phases present at the stage when the liquid contains 10 wt.% Ag and lies on the $L \rightarrow Cu + Cu_3P$ eutectic valley; and
iv. the percentages of primary phase and of binary and ternary eutectic mixtures respectively present at room temperature. (Assume equilibrium conditions and neglect solid solubility).

b. Sketch isothermal sections through the system at

Table 5.2

Temperature (°C)	Composition (wt.%)		
	A	B	C
1100	-	35	65
1000	70	30	-
950	45	20	35
800	50	-	50

Table 5.3

Solid Solution	Solid Solubility Limits (wt.%)	
	At 945°C	At Room Temperature
α	50A, 40B, 10C	60A, 35B, 5C
β	12A, 80B, 8C	3A, 95B, 2C
γ	10A, 45B, 45C	5A, 40B, 55C

i.   800°C and

ii.   at a temperature above 646°C but below the binary eutectic temperatures. (Make assumptions as necessary concerning the liquidus isotherms and neglect solid solubility).

5.   Three metals A, B and C show complete liquid solubility in one another and partial solid solubility. The ternary system contains only three solid phases namely α, β and γ solid solutions. The melting points of A, B and C are 900, 1200 and 1000°C, respectively.

The system contains invariant reactions at the temperatures and compositions shown in Table 5.2.

Solid solubility data are shown in Table 5.3.

a.   Draw a liquidus projection of the system and an isothermal section at room temperature, consistent with the above data.

b.   State the nature of each invariant reaction in the system.

c.   Describe the equilibrium solidification sequence of an alloy containing 30% A, 55% B, 15% C, and calculate the relative amounts of the phases present in this alloy at 945°C.

d.   Describe the equilibrium solidification sequence of an alloy containing 20% A, 35% B, 45% C.

6. Systems Containing More than Four Phases

6.1 INTRODUCTION

The presence of one or more intermediate phases (either binary or ternary) in addition to three primary solid solutions (or pure components) leads to many possible forms of ternary system. Several examples are discussed, but the space models depict only the liquidus features and not the phase regions within the models. Examples of ternary systems of industrial importance, including intermediate phases, are discussed as case studies in Chapter 8.

6.2 AN INTERMEDIATE PHASE STABLE UP TO ITS MELTING POINT (I.E. CONGRUENTLY MELTING) EXISTS IN ONE OF THE BINARY SYSTEMS

In the system ABC depicted in Figure 6.1, it is assumed that system AB contains an intermediate phase, X, that is stable up to its melting point, while systems AC and BC are eutectic in nature. Four binary eutectic curves enter the ternary system from the binary systems, and assuming that there is no solid solubility of the components, the solidification reactions are as follows:

$DI : L \rightarrow A + X$

$EG : L \rightarrow B + X$

$JI : L \rightarrow A + C$

$FG : L \rightarrow B + C$

Within the ternary system there is also a curve IHG representing the eutectic reaction $L \rightarrow X + C$. As shown in Figure 6.1 this curve is in the form of a 'saddle' with a maximum point at H.

The system contains two invariant eutectic reactions at points I and G:

At $I : L \rightarrow A + X + C$

At $G : L \rightarrow B + X + C$

The regions of primary solidification of A, B, C and X shown in the projected view of Figure 6.2, are $Adij$, $Begf$, $Cjihgf$ and $deghi$ respectively. The line XC divides the system into two triangular portions AXC and BXC, each of which is a ternary eutectic type of

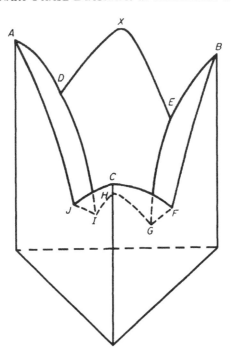

Fig. 6.1 Space model of system containing an intermediate phase X, stable up to its melting point, and showing two invariant reactions $L \rightarrow A + X + C$ and $L \rightarrow B + X + C$ (Only liquidus features are shown).

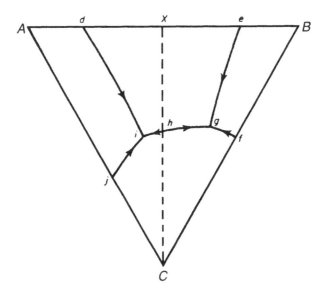

Fig. 6.2 Projected view of system of Figure 6.1.

system with no solid solubility of the components: the components in these two portions are A, X and C, and B, X and C respectively.

XC can be termed a 'quasi-binary' section or system; it has the characteristics of a binary system in that during the course of solidification of any alloy lying in it, the compositions of liquid and solid remain in the section; i.e. all tie-lines, and the three-phase equilibrium line all lie in the plane of the section.

The form of section XC (shown in Figure 6.5) is eutectic in type, and the eutectic point for the reaction $L \rightarrow X + C$, corresponds to the saddle-point, H, in Figure 6.1.

If the ternary system is assumed to show partial solid solubility of the components, it may be represented in a projected view as in Figure 6.3, while an isothermal section in the solid state is shown in Figure 6.4.

When there is partial solid solubility as illustrated, the section joining C to the maximum point X in the binary system AB is not necessarily quasi-binary in the exact sense defined above. For example, the section might be as shown in Figure 6.6, since the 'saddle top' of the $L + \delta + \gamma$ region does not necessarily lie in the plane XC.[5] However, in practice, the term quasi-binary is commonly applied to cases where some degree of solid solubility exists, and if the extent of the solubility is small there is likely to be little deviation from the strictly quasi-binary case.

The solidification sequences of typical ternary alloys will not be discussed here; the principles previously described for ternary eutectic systems are applicable.

With the same types of binary system as in the case described above, it is possible for the ternary system to contain one invariant peritectic reaction and one invariant eutectic reaction, instead of containing two invariant eutectics. This alternative case is illustrated in Figure 6.7 and in the projected view shown in Figure 6.8; the regions of primary solidification of A, B, C and X are $Adhi$, $Begf$, $Cihgf$ and $degh$ respectively. Assuming that there is no solid solubility, point H represents the composition of liquid taking part in the reaction.

$$L + A \rightarrow X + C$$

The peritectic reaction arises because the intersection of the curves DH ($L \rightarrow A + X$) and IH ($L \rightarrow A + C$) occurs at a point H which lies outside the triangle formed by the compositions of the solid phases A, X and C. Curve HG representing the reaction $L \rightarrow X + C$ runs to lower temperatures from H, and meets the eutectic curves EG ($L \rightarrow B + X$) and FG ($L \rightarrow B + C$) to give rise to the invariant reaction at G ($L \rightarrow B + X + C$).

In the solid state, the system is divided into the regions AXC and BXC as in the first case discussed. However, section XC is not quasi-binary; consideration of the solidification of alloys lying in this section will show that the compositions of the liquid and solids do not remain in the plane of the section throughout the whole course of solidification, although the final structure under equilibrium conditions consists of X and C.

All alloys lying within the region $AXhC$ (Figure 6.8) undergo the ternary peritectic reaction during solidification, the liquid compositions reaching point h, either via curve ih or curve dh. For alloys in the triangle AXC, all the liquid is consumed in the reaction under equilibrium conditions. If the alloy composition lies in the triangle XhC, all of component A is consumed, and the remaining liquid deposits X and C by a eutectic reaction along curve hg as cooling proceeds, until the liquid composition reaches g when the reaction $L \rightarrow B + X + C$ completes the solidification process. If the alloy composition

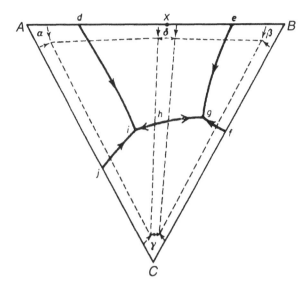

Fig. 6.3 Projected view of Figure 6.1 assuming partial solid solubility of the components.

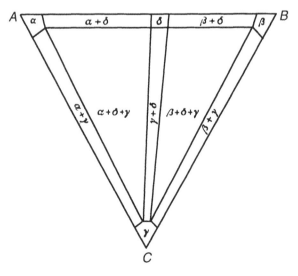

Fig. 6.4 Isothermal section in the solid state for the system of Figure 6.1 showing partial solid solubility.

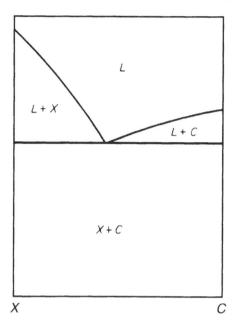

Fig. 6.5 Quasi-binary section *XC* in the system of Figure 6.1.

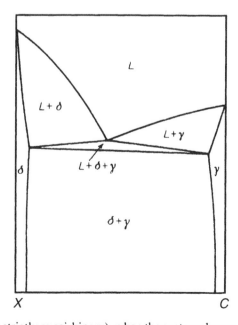

Fig. 6.6 Section *XC* (not strictly quasi-binary), when the system shows partial solid solubility.

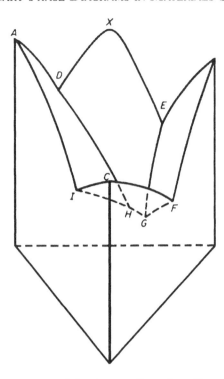

Fig. 6.7 Space model of system containing intermediate phase X, and showing two invariant reactions $L + A \rightarrow X + C$ and $L \rightarrow B + X + C$. (Only liquidus features are shown).

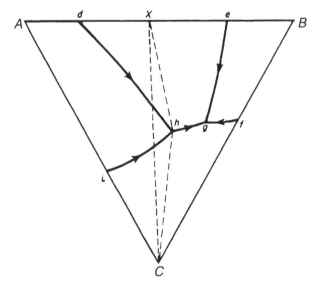

Fig. 6.8 Projected view of system of Figure 6.7.

lies on the line XC then both liquid and A are consumed in the peritectic reaction leaving a mixture of X and C.

Figure 6.9 shows a projected view of the system assuming that partial solid solubility exists.

6.3 AN INTERMEDIATE PHASE FORMED BY A PERITECTIC REACTION EXISTS IN ONE OF THE BINARY SYSTEMS

In the system ABC shown in Figures 6.10 and 6.11, in which it is assumed that there is no solid solubility of the components, systems AC and BC are eutectic in nature, while system AB contains an intermediate phase, X, formed by the reaction $L + A \rightarrow X$ (see Figure 6.12); in the ternary system the peritectic curve DH ($L + A \rightarrow X$) intersects the eutectic curve IH ($L \rightarrow A + C$) at the invariant peritectic point H ($L + A \rightarrow X + C$). Curve HG representing the reaction $L \rightarrow X + C$ intersects the eutectic curves EG ($L \rightarrow X + B$) and FG ($L \rightarrow B + C$) to give the reaction $L \rightarrow B + X + C$ at point G. The portion of the liquidus surface $DEGH$ represents the primary separation of the intermediate phase X.

In the solid state the system is divided into the two regions AXC and BXC as in the case previously discussed, and section XC is not quasi-binary. All alloys within the region $AXhC$ (Figure 6.11) undergo the peritectic reaction and, depending on whether an alloy lies within the triangle AXC or XhC, either the liquid or component A is consumed, as discussed for the previous case. The solidification sequence of alloys in the region Xdh is as follows: the primary separation of A causes the liquid composition to reach curve dh when the reaction $L + A \rightarrow X$ occurs; all the A is consumed and the liquid composition then leaves curve dh and changes on a path on the surface $degh$ as X is deposited, until the liquid composition reaches either hg or eg, and then finally attains point g. Figure 6.13 shows a projected view of the system assuming partial solid solubility of the components.

It is possible that the curve dh representing the $L + \alpha \rightarrow \delta$ peritectic reaction may change direction as it moves into the ternary system, and thus undergo a transition to a eutectic reaction, $L \rightarrow \alpha + \delta$, as discussed in Chapter 4 (Figures 4.17 and 4.19). In this case (Figure 6.14) the junction of the $L \rightarrow \alpha + \delta$ curve with the $L \rightarrow \alpha + \gamma$ curve, ih, within the triangle formed by the compositions of the three solids α, δ and γ will lead to an invariant eutectic reaction $L \rightarrow \alpha + \delta + \gamma$. The system then contains two invariant eutectics, connected by a saddle-type eutectic curve $L \rightarrow \delta + \gamma$ (Figure 6.14), but section XC is not quasi-binary (see also Figure 6.15 and page 193).

6.4 MORE THAN ONE BINARY INTERMEDIATE PHASE EXISTS IN THE SYSTEM

Figures 6.16a and b illustrate two possible arrangements of quasi-binary sections in a ternary system ABC in which systems AB and BC each contain a congruently-melting intermediate phase, designated X and Y respectively.

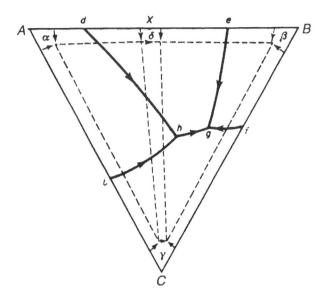

Fig. 6.9 Projected view of system of Figure 6.7, assuming partial solid solubility of the components.

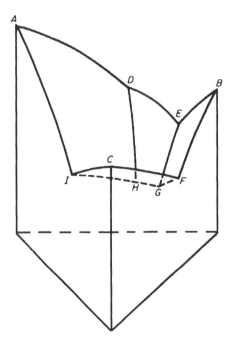

Fig. 6.10 Space model of system containing an intermediate phase X (formed by a peritectic reaction), and containing two invariant reactions $L + A \rightarrow X + C$ and $L \rightarrow B + X + C$. (Only liquidus features are shown).

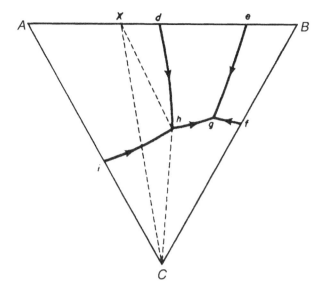

Fig. 6.11 Projected view of system of Figure 6.10.

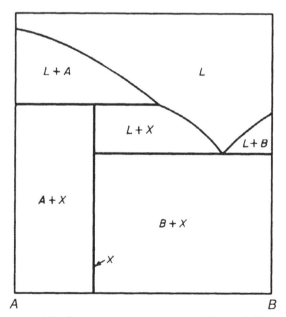

Fig. 6.12 Binary system *AB* of ternary system shown in Figure 6.10.

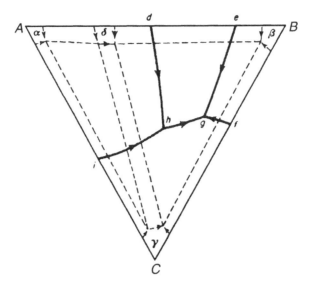

Fig. 6.13 Projected view of system of Figure 6.10 assuming partial solid solubility.

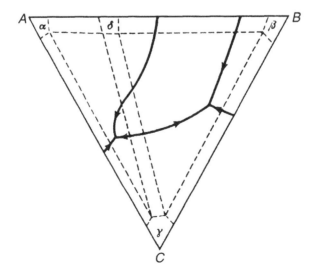

Fig. 6.14 Projected view of system in which the binary peritectic reaction $L + \alpha \rightarrow \delta$ changes to a eutectic reaction $L \rightarrow \alpha + \delta$.

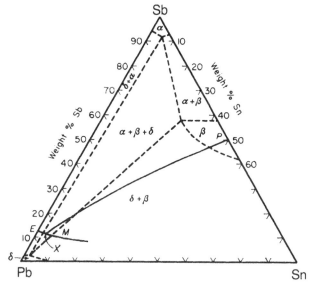

Fig. 6.15 Liquidus projection (shown as full lines) of part of the Pb–Sb–Sn system, together with part of a solid state isothermal section (dashed lines) for the system. α and δ are primary solid solutions based on Sb and Pb respectively, and β represents the solid solution based on the intermediate phase, SbSn. The reactions represented on the liquidus projection are as follows: *EX*: $L \rightarrow \alpha + \delta$; *MX*: $L \rightarrow \beta + \delta$; *PX*: $L + \alpha \rightarrow \beta$; *X*: $L \rightarrow \alpha + \beta + \delta$; (From Metals Handbook, 1948, Reproduced with permission from *ASM International*, Materials Park, OH 44073-0002 U.S.A).

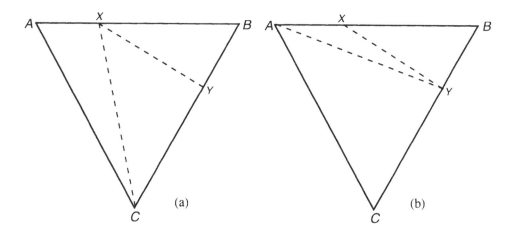

Fig. 6.16 Possible arrangements of quasi-binary sections in a system containing two binary intermediate phases.

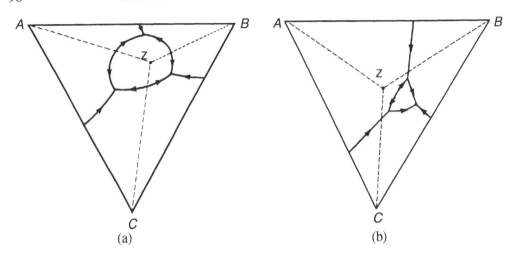

Fig. 6.17 Projected view of a system containing a ternary intermediate phase: (a) congruently melting and (b) formed by a peritectic reaction $L + A \leftrightarrow Z$.

In Figure 6.16a XY and XC represent quasi-binary sections, while in Figure 6.16b, XY and YA are quasi-binary. In both cases, the ternary system is divided into three parts, each of which may be considered as an 'independent' ternary system. It follows from the nature of quasi-binary sections that they cannot cross one another, and thus it is not possible for XC and YA to be the pair of quasi-binary sections.

To determine experimentally the particular pair of quasi-binary sections in an actual system (if in fact such exist), an alloy may be made at the intersection of the lines XC and YA and after being annealed to equilibrium, it may be examined to find what phases are present. If only phases X and C are found, then XC is quasi-binary, and if A and Y are found, then YA is the quasi-binary section, assuming that peritectic reactions are not involved.

6.5 TERNARY INTERMEDIATE PHASES ARE PRESENT

Concerning systems containing one or more ternary intermediate phases, hypothetical examples are illustrated in Figure 6.17. Figure 6.17a represents a system containing a congruently-melting ternary phase, Z, the binary systems each being of the eutectic type. Three quasi-binary sections, AZ, BZ, and CZ exist, and the system is divided into three parts. In Figure 6.17b, the binary systems are of the eutectic type, but the ternary intermediate phase forms by a peritectic reaction: $L + A \rightarrow Z$. The liquidus shows two invariant peritectic reactions: $L + A \rightarrow B + Z$ and $L + A \rightarrow C + Z$ respectively. The third ternary invariant reaction is $L \rightarrow B + C + Z$.

Rules relating the maximum number of independent ternary systems into which a ternary system can be divided when binary and/or ternary intermediate phases occur are discussed and illustrated by Rhines,[6] who also illustrates typical systems.

Examples of actual systems containing both binary and ternary intermediate phases are common, e.g. in metallic and ceramic systems. Figures 6.18a and b, illustrates part of the Al–Cu–Mg system and Figures 6.19a and b depicts a complex oxide system.

6.6 PROBLEMS

1. A ternary system ABC shows complete liquid solubility and partial solid solubility, forming primary solid solutions α, β and γ, based on A, B and C, respectively. System BC contains a peritectic reaction (liquid $+ \beta \rightarrow \delta$) at 800°C, in which δ is an intermediate phase containing 40% C, and the liquid contains 50% C. The ternary system contains two invariant reactions as shown in Table 1.

Table 1

Temperature	Liquid Composition	Solid Phases Involved
650°C	30%A, 32%B, 38%C	α, β, δ
620°C	37%A, 8%B, 55%C	α, δ, γ

The solid solubility ranges of α, β, γ and δ are very small (<1%).
a. Draw:
   i. a liquidus projection for the system consistent with the data, and state the nature of the reactions shown,
   ii. an isothermal section corresponding to a temperature of 600°C.
b. By reference to the liquidus projection and the 600°C isothermal describe a possible solidification sequence for an alloy containing 10% A, 60% B and 30% C, assuming equilibrium cooling conditions. Calculate the equilibrium proportions of β phase present in this alloy at 651 and 649°C, respectively.
2. With reference to Figure 6.15 describe the equilibrium solidification sequences of the following alloys:
   a. Pb–15%Sb–5%Sn;
   b. Pb–15%Sb–10%Sn,
3. a. With reference to Figures 6.18a and b describe the equilibrium solidification sequences of the following Al-based alloys:

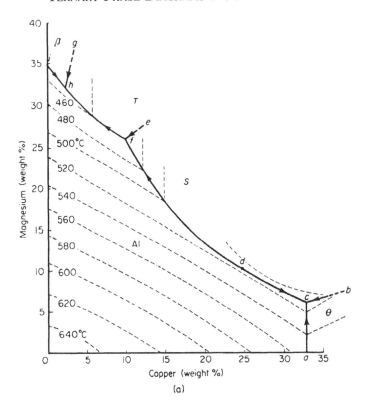

Fig. 6.18 (a) Liquidus projection of part of the Al–Cu–Mg system. The phases involved are as follows: aluminium solid solution (Al); $CuAl_2$(θ): Mg_2Al_3(β); a ternary intermediate phase corresponding approximately to Al_2CuMg(S); and a ternary intermediate phase corresponding approximately to Al_6CuMg_4(T). The fields of primary phase separation are labelled and some isotherms are shown. The reactions represented are as follows.

ac : $L \rightarrow Al + \theta$ f : $L + S \rightarrow Al + T$

bc : $L \rightarrow \theta + S$ fh : $L \rightarrow Al + T$

c   : $L \rightarrow Al + \theta + S$ gh : $L \rightarrow \beta + T$

cdf : $L \rightarrow Al + S$ ih : $L \rightarrow S + Al + \beta$

ef : $L + S \rightarrow T$ h : $L \rightarrow Al + \beta + T$

The section Al–S is approximately quasi-binary, and d is the saddle point on curve cf.

  i.   3%Cu,15%Mg;

 ii.   6%Cu, 12%Mg;

iii.   24%Cu, 10%Mg:

 iv.   4.5%Cu, 2.5%Mg.

  b.   Samples of an Al-4.5%Cu-2.5%Mg alloy and of a binary Al-4.5%Cu alloy are solution treated at 500°C for several hours, water-quenched and then heated at

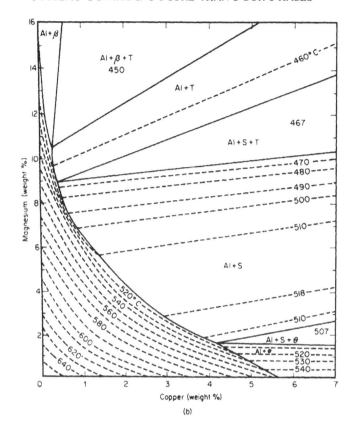

Fig. 6.18 (b) Solidus projection of part of Al-Cu-Mg system; (Figures 6.18 a and b are reproduced from H. W. L. Phillips, Information Bulletin 25, *Equilibrium Diagrams of Aluminium Alloy Systems,*[23] 1961, with the permission of the Aluminium Federation, London).

200°C for 24 hours. From the data in Figure 6.18b and taking the solubilities of Cu in Al and Mg in Al at 200°C as ~0.1 wt.% and ~3 wt.%, respectively, comment on the difference in the heat treatment response of these two alloys.

4. With reference to the system SiO_2–$Li_2O \cdot SiO_2$–$Li_2O \cdot Al_2O_3 \cdot 4SiO_2$:

a. Draw an isothermal section for a temperature of 970°C. Make use of the following information in addition to that shown in Figure 6.19. In the spodumene-SiO_2 binary system, β-spodumene shows extensive solid solubility for SiO_2. β-spodumene co-exists with $Li_2O \cdot SiO_2$ over the range 0–23% SiO_2 and with $Li_2O \cdot 2SiO_2$ over the range 23–33% SiO_2. Assume negligible solid solubility in the following compounds: $Li_2O \cdot SiO_2$, $Li_2O \cdot 2SiO_2$, and SiO_2.

b. Draw a graph showing how the percentage of liquid present in equilibrium at 1100°C varies with composition along a line joining SiO_2 to a binary composition of 20% spodumene–80% $Li_2O \cdot SiO_2$.

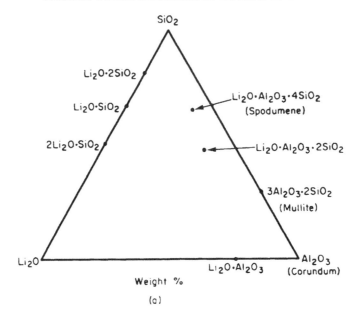

(a)

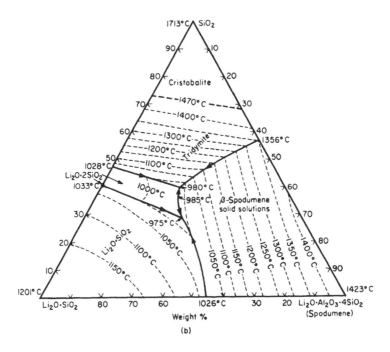

(b)

Fig. 6.19 (a) Diagram showing the binary and ternary phases present in the ternary system Li_2O-SiO_2-Al_2O_3 and (b) Liquidus projection for the part of the above system SiO_2-$Li_2O\cdot SiO_2$-$Li_2O\cdot Al_2O_3\cdot 4SiO_2$, the composition scales being plotted taking these phases as the components. (Reprinted with permission of The American Ceramic Society, Post Office 6136, Westerville, Ohio 43086-6136, Copyright 1956 by The American Ceramic Society. All rights reserved).

7. Principles of Using Liquidus Projections for Equilibrium and Non-Equilibrium Solidification in Metallic and Ceramic Systems

7.1 INTRODUCTION

Equilibrium data presented as liquidus projections are often used to study solidification sequences. In order to make the fullest use of such projections, information is also required on the solid state constitution of a given system at the completion of solidification.

Two ways of presenting the data on solid state constitution may be considered, namely, a solidus projection, and an isothermal section at a temperature below the lowest solidus temperature in the system. In general, in systems with no solid solubility (e.g. as is effectively the case in many ceramic systems), these two modes of presentation are equivalent for the purpose required. In such systems the solidification of all ternary alloys is completed by an invariant reaction, and the solidus is composed of a plane, or planes, depending on the number of invariant reactions contained in the system. The solidus projection and the isothermal section show one or more three-phase regions, bounded by lines joining pairs of phases co-existing in equilibrium. In systems with solid solubility, the solidus projection is the most accurate and informative means of presenting the required data, especially if isotherms are depicted. If significant solid solubility changes or phase transformations occur in the solid state, the isothermal section at a temperature just below the lowest solidus temperature may not be a satisfactory source of information for studying the complete cooling sequences of various alloys. Further, in such cases, isothermal sections at various temperatures through the solid state region would be needed if the changes occurring during the cooling of given alloys to low temperatures were to be traced accurately.

The preceding chapters have included accounts of the procedures involved in studying solidification sequences by reference to liquidus and solidus projections: various examples have been considered, some of them relatively complex. In this chapter, a summary is given of the basic principles and procedures; it is hoped that this will assist the reader in reviewing the examples previously discussed and in the consideration of other systems.

7.2 PRIMARY PHASE FIELDS ON A LIQUIDUS PROJECTION

The liquidus projection for a system is commonly labelled to show the regions of primary solidification. The location of a given alloy composition within a particular phase field on the projection serves to indicate the primary phase that separates during solidification. If isotherms are shown on the liquidus projection, they can be used to determine the liquidus temperature for a given alloy. As solidification proceeds by separation of primary phase the composition of the liquid changes progressively. This change is represented by a path on the liquidus projection, and occurs in a direction representing a lowering of temperature on the liquidus surface. This path is formed by the extremities of the tie-lines representing the liquid compositions in equilibrium with the primary phase at successively lower temperatures.

When the primary phase is of fixed composition, e.g. a pure metal, or an intermediate phase showing no solid solubility, the path is simply the extension of the straight line joining the composition of the primary phase and the alloy composition. At any temperature during the primary solidification stage, if the liquid composition is known, the amounts of liquid and solid can be calculated by applying the lever rule to the tie-line. Thus the tie-line AXZ in Fig. 5.4, represents the equilibrium at the end of the primary solidification stage when the liquid composition has just reached a point on the eutectic curve dg. From this tie-line the amount of primary A is represented by the ratio XZ/AZ.

In systems where the primary phase is a solid solution, its composition changes during solidification. For these cases the accurate tracing of the paths along which the liquid and solid compositions change requires experimental or thermodynamically calculated data regarding tie-lines; these data are often not available. However, when the extent of the solid solubility is small, the procedure for a phase showing no solid solubility may be used as an approximation, i.e. taking the extension of the straight line joining the composition of the pure metal or intermediate phase. With extensive solid solubility, however, the adoption of this procedure may be quite misleading. It is then preferable to attempt to estimate the change in liquid and solid compositions with reference to the solidus projection, which will at least help in locating the likely compositions of liquid and solid co-existing at the beginning of a three-phase reaction immediately following the primary solidification.

It should be noted that where solid solubility exists it is not possible to calculate exactly the relative amounts of liquid and solid co-existing during primary solidification unless tie-line data are available.

7.3 CURVES ON A LIQUIDUS PROJECTION

A curve such as, for example, a eutectic valley, represents the path along which the liquid composition changes while undergoing a three-phase solidification reaction; the liquid co-exists with two solids in a eutectic or peritectic reaction, and with a solid and another liquid in a monotectic reaction. The three-phase reaction commences when the path representing the changing liquid composition during the primary stage intersects a curve on the liquidus projection.

The three-phase reaction then proceeds as the liquid composition changes along the curve; commonly, arrows are shown on the curves to indicate the direction of decreasing temperature. With regard to the solid phases separating during the three-phase reaction, their compositions will be fixed in the case of pure metals or other components showing no solid solubility. When the phases separating are solid solutions, their compositions change along the relevant solid solubility curves; detailed experimental or thermodynamically calculated data would be required to locate the tie-triangles which would be necessary for a calculation of the amounts of liquid and solid existing at the various temperatures, using the centre-of-gravity rule.

Reference to the limits of the solid state regions (as shown on a solidus projection or isothermal section), will show if the solidification of a given alloy is completed by the three-phase reaction to give a two-phase alloy (e.g. as in alloy X of Figure 5.3). Alternatively, the liquid may undergo an invariant reaction as a sequel to the three-phase reaction, (e.g. as in alloy Y of Figure 5.3 which consists of $\alpha + \beta + \gamma$ when solid, and which therefore must complete its solidification by the ternary eutectic at g).

7.4 POINTS ON LIQUIDUS PROJECTIONS

Points formed by the intersection of curves on a liquidus projection represent the composition of liquid involved in an invariant reaction. Such a point is formed by the intersection of three liquidus curves, and when the liquid composition reaches it, the invariant reaction occurs. Assuming equilibrium conditions, the temperature remains constant until the reaction is completed, and during the reaction the compositions of all four phases involved remain unchanged.

The nature of a ternary invariant reaction represented by a point can be considered by reference to the arrows shown on the intersecting liquidus curves and also to the number of three phase equilibria above and below the invariant temperature, (Ch. 5.1) according to the classification of Rhines (6), and summarised in the following Table:

Invariant Reaction Class	Number of Three-Phase Equilibria Above and Below the Invariant Reaction Temperature	
	Above	Below
I	3	1
II	2	2
III	1	3

Consider the following liquidus projection situations, involving a liquid (not associated with a miscibility gap) and three solid phases in relation to solidification:

(i)

This configuration can only represent a Class I reaction, $L \rightarrow \alpha + \beta + \gamma$, since there are three three-phase equilibria descending towards the ternary invariant, and one below it.*

(ii)

This configuration could represent a class II reaction, such as $L + \alpha \rightarrow \beta + \gamma$, involving two three-phase equilibria above the invariant e.g. $L \rightarrow \alpha + \beta$ and $L + \alpha \rightarrow \gamma$, giving rise to the four-phase reaction: $L + \alpha \rightarrow \beta + \gamma$; below the invariant temperature the three phase regions (tie-triangles) would be $L + \beta + \gamma$ and $\alpha + \beta + \gamma$. However, as discussed below, without further information on the nature of all of the four three-phase regions associated with the invariant plane, the reaction depicted might be of the Class I type, $\alpha \rightarrow \beta + \gamma + L$.

(iii)

This configuration could represent a Class III reaction, such as $L + \alpha + \beta \rightarrow \gamma$, with one three-phase equilibrium above the invariant temperature e.g. $L \rightarrow \alpha + \beta$, and three three-phase equilibria below the invariant involving: $\alpha + \beta + \gamma$, $L + \alpha + \gamma$, and $L + \beta + \gamma$ respectively. However, as mentioned in (ii), information is required on the nature of the four three-phase regions in order to define uniquely the nature of the reaction depicted; the situation shown in (iii) could represent a Class II invariant: $\alpha + \beta \rightarrow \gamma + L$.

As indicated above concerning the situations shown in (ii) and (iii), the information provided by the liquidus projection does not uniquely define the Class of a four phase equilibrium. This is the case when a ternary system involves more than three solid phases, and when solid state transformations, such as the eutectoidal transformation of an intermediate phase or allotropic transformations occur. The possibility then exists of a three phase field involving three solid phases participating with fields containing liquid + two solid phases to produce an invariant reaction. It is then necessary to know the nature of the phase fields above and below the invariant plane in order to deduce the Class of the invariant.

---

*Ricci[4] notes a feature concerning the configuration of the projections of three univariant curves associated with three-phase reactions meeting at an invariant point. At the intersection point of the curves there is no angle > 180° between adjacent curves.

Figures 7.1 and 2 illustrate situations involving liquidus features such as those shown in ii and iii above which could give rise to Class I[6] and Class II reactions respectively.

Considering again (ii) and (iii) above as examples in the situation where they represent the peritectic types of reaction ($L + \alpha \rightarrow \beta + \gamma$ (Class II) and $L + \alpha + \beta \rightarrow \gamma$ (Class III), reference to the solid state constitution will reveal whether a given alloy composition lies in the three-phase $\alpha + \beta + \gamma$ region, in which case solidification is completed by the invariant reaction. If the composition of an alloy that undergoes the invariant reaction lies outside the $\alpha + \beta + \gamma$ region, liquid will remain at the completion of the invariant reaction (see for example region *ngo* of Figure 5.11). This liquid will then undergo a three-phase reaction along a curve running from the invariant point to lower temperatures, and assuming that the system does not contain any other invariant reactions the alloy will solidify as a two-phase mixture, or possibly, for the $L + \alpha + \beta \rightarrow \gamma$ reaction, a single phase γ structure may result (e.g. as in region *ogp* of Figure 5.15).

If the system contains another invariant reaction, involving an additional solid phase, say δ, the liquid composition may reach the second invariant point before solidification is complete; reference to the solid state constitution will show whether this will be so, e.g. see Figure 6.9.

7.5 NON-EQUILIBRIUM SOLIDIFICATION

In normal practice, casting processes do not permit equilibrium to be attained during solidification. This is because significant diffusion in the solid phases formed during solidification is required for equilibrium conditions to be met. More usually, casting microstructures exhibit 'coring' where there is a composition gradient across dendrite arms. When subject to even faster cooling rates, e.g. $> 10^3$ K s^{-1} even more pronounced non-equilibrium effects can occur; for example extensive undercooling, suppression of equilibrium phases, glass formation etc. Some examples are briefly considered here. The occurrence of coring can commonly produce a three-phase structure, where a two-phase structure would exist under equilibrium. Thus, alloy X of Figure 5.3 might contain some γ phase, due to the liquid composition reaching point g under non-equilibrium cooling conditions. The effects of non-equilibrium cooling may be particularly marked in systems containing peritectic reactions, due to the reactions not proceeding to completion. For example, in the system represented in Figure 6.9, alloys lying in the $\alpha + \delta + \gamma$ region when in equilibrium, might contain some β phase after rapid cooling, due to the invariant peritectic reaction at h failing to go to completion. The composition of the remaining liquid will then eventually reach point g to undergo the ternary eutectic reaction.

The aluminium–copper–magnesium system depicted in Figures 6.18a and b illustrates this type of situation for the non-equilibrium cooling of certain alloys. For example, an alloy containing ~15%Cu and 17%Mg should complete its solidification by the invariant peritectic reaction at point f to give Al + T + S; however, with rapid cooling the liquid

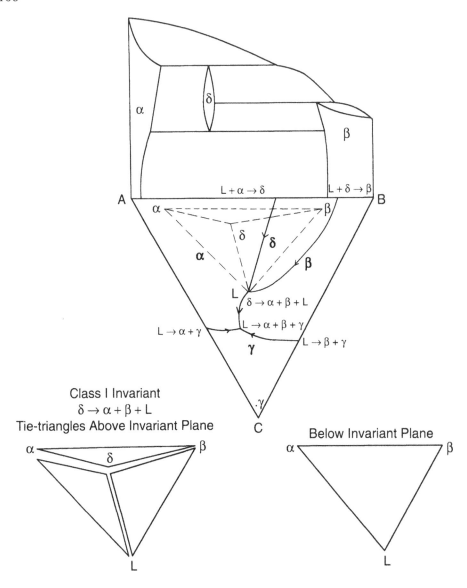

Fig. 7.1 The system *ABC* shows primary solid solutions α, β and γ, based on *A*, *B* and *C* respectively. δ is an intermediate phase formed in system *AB* by a peritectic reaction. δ transforms eutectoidally as shown to α + β. Systems *AC* and *BC* each show a eutectic reaction The ternary system shows an invariant reaction involving δ, α, β and liquid. Consideration of the associated three phase equilibria shows that the tie triangle (δ + α + β) representing the eutectoid, moves to lower temperature and at the invariant plane it meets the three phase regions originating from the peritectic reactions (*L* + α → δ) and (*L* + δ → β). The convergence of these three regions at the invariant temperature produces the reaction : δ → α + β + *L*. Below the invariant plane there is one three-phase region: *L* + α + β. Thus the invariant reaction is Class I, although it shows the liquidus configuration of (ii). The other invariant reaction shown in Figure 7.1 represents a eutectic : *L* → α + β + γ (Class I).

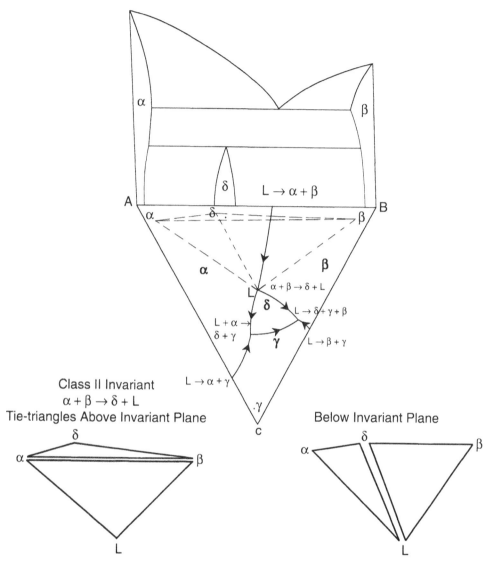

Fig. 7.2 The primary solid solutions are the same as in Fig. 7.1 and the binary systems *AC* and *BC* also contain the same eutectic reactions. The system *AB* however, shows a eutectic, and the δ phase forms peritectoidally in contrast to the eutectoid in Fig. 7.1. The ternary system shows a ternary invariant reaction involving :liquid, δ, α and β. The tie-triangle representing the peritectoid reaction moves to lower temperature and converges with the *L* + α + β region associated with the binary eutectic reaction. This convergence gives rise to a ternary invariant reaction α + β → δ + *L*, which produces two three-phase regions: *L* + α → δ and *L* + β → δ. The presence of two three-phase regions above the invariant plane and two below the plane corresponds to a Class II reaction, although the configuration of liquidus curves is the same as that shown in (iii). Considering the other two ternary invariant reactions in Fig. 7.2, one of these shown as peritectic (Class II) : *L* + α → γ + δ, and the other as eutectic *L* → δ + γ + β respectively. (This latter reaction could be shown as a ternary peritectic: *L* + δ → γ + β by locating the liquid composition at the invariant point outside the δ + γ + β triangle).

composition can reach the ternary eutectic point h representing the reaction: $L \rightarrow Al + \beta + T$. This means that, in addition to Al, T, and S, the solid alloy will contain some β phase.

As a further example consider the effect of rapid solidification completely suppressing the formation of an intermediate phase, and giving rise to a metastable reaction, as depicted in Figure 7.3[24] The phases and phase relationships in the system are essentially the same as that considered in Figure 6.9.

The system ABC shows partial solid solubility of the three components, forming solid solutions α, β and γ respectively. The binary systems AC and BC are of the eutectic type: $L \rightarrow \alpha + \gamma$ and $L \rightarrow \beta + \gamma$ respectively. System AB contains a congruently melting intermediate phase, δ, showing a range of solid solubility, and associated with the eutectic reactions: $L \rightarrow \alpha + \delta$ and $L \rightarrow \beta + \delta$.

If the assumption is made that the δ phase is difficult to nucleate during cooling, then rapid solidification of alloys of composition lying in the primary δ region of the liquidus can give rise to a metastable reaction, as illustrated in Figure 7.3. During rapid solidification the primary δ region of the stable system is eliminated as are the two ternary invariant reactions: $L + \beta \rightarrow \delta + \gamma$ and $L \rightarrow \alpha + \beta + \delta$ and the associated binary eutectic reactions: $L \rightarrow \beta + \delta$ and $L \rightarrow \delta + \gamma$. The metastable situation involves a ternary eutectic $L \rightarrow \alpha + \beta + \gamma$, formed by the convergence of the metastable extensions of the $L \rightarrow \alpha + \gamma$ and $L \rightarrow \beta + \gamma$ with a metastable eutectic: $L \rightarrow \alpha + \beta$, originating from the binary AB system. This situation could favour glass formation in the ternary system.

7.6 QUANTITATIVE APPROACH TO NON-EQUILIBRIUM SOLIDIFICATION

Particularly in recent years, following earlier work by Gulliver,[25] Scheil[26] and Pfann[27] there has been great interest in the quantitative predictions of solidification paths. The work of these authors utilised the assumption that solute diffusion in the solid state was very small, so much so that it could be considered to have negligible effect on solidification. On the other hand, diffusion in the liquid state was extremely fast, fast enough so that it could be assumed that diffusion was complete. In this context, Scheil developed a mathematical approach that would describe solidification under these conditions, and presented a formula that calculated the amount of solid transformed as a function of temperature.

Consider a liquidus-solidus region, assuming linear liquidus and solidus lines (i.e. constant values of the partition coefficient, k), and apply the lever rule to an alloy of original composition C_o. For equilibrium conditions the composition of the solid, C_s, as a function of the fraction solid transformed, f_s, is given by:

$$C_s = kC_o/(f_s(k-1) + 1) \tag{i}$$

This equation can also be expressed as follows to represent the fraction of solid as:

$$f_s = [1/(1-k)][(T_L-T)/T_f-T] \tag{ii}$$

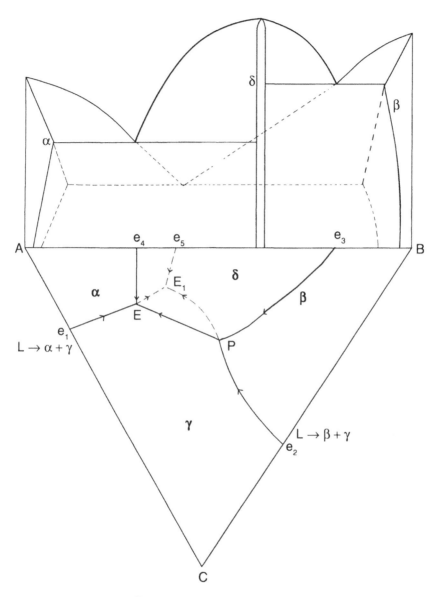

Fig. 7.3 Liquidus projection[24] for a ternary system containing an intermediate phase δ, which has a congruent melting point in the binary system AB, and which also forms during a peritectic reaction in the ternary system: $L + \beta \rightarrow \gamma + \delta$. This ternary peritectic is suppressed by rapid solidification, leading to a metastable eutectic reaction: $L \rightarrow \alpha + \beta + \gamma$.
Invariant reactions:

e_1 : $L \rightarrow \alpha + \gamma$ e_2 : $L \rightarrow \beta + \gamma$ e_3 : $L \rightarrow \beta + \delta$
e_4 : $L \rightarrow \alpha + \delta$ e_5(metastable) : $L \rightarrow \alpha + \beta$
P : $L + \beta \rightarrow \gamma + \delta$ E : $L \rightarrow \alpha + \gamma + \delta$ E_1(metastable):$L \rightarrow \alpha + \beta + \gamma$.

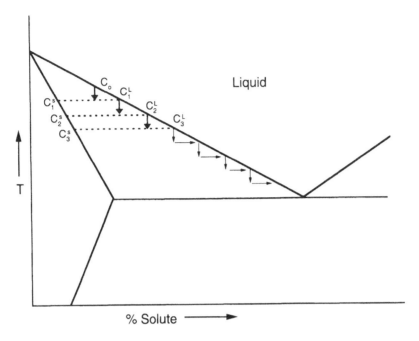

Fig. 7.4 Schematic representation of solidification under Scheil conditions. (Reprinted from N. Saunders and A. P. Miodownik, *CALPHAD*: *Calculation of Phase Diagrams*, 1996, p.443, with permission from Elsevier Science).

where T_L and T_f are the equilibrium liquidus and solidus temperatures for the alloy.

Consider a non-equilibrium situation, assuming that the solidification rate is sufficiently rapid that solid state diffusion is negligible, while diffusion in the liquid is fast enough to be complete, then eqn. (i) can be rewritten as

$$C_s = kC_o(l\text{-}f_s)^{k\text{-}1} \tag{iii}$$

And eqn. (ii) as

$$f_s = 1 - [(T_f\text{-}T)/T_f\text{-}T_L] \tag{iv}$$

The latter equation, often called the Scheil equation, has been used quite extensively to describe solidification under non-equilibrium cooling conditions. However, eqns. (i-iv) cannot be derived by the same mathematical procedure if k varies with temperature and/or composition, which is the more usual case in practice. Furthermore, while the Scheil equation can be applied to dendritic solidification, it cannot be applied to subsequent eutectic formation. The CALPHAD procedure enables these limitations to be overcome, and also allows multi-component systems to be dealt with (see Chapter 8).

In terms of the development of a 'cored' structure in the solidified crystals, which can be typically dendritic in morphology, the basis of the procedure as applied to a simple binary system is illustrated by reference to Figure 7.4. When liquid, C_0 is cooled slightly below the liquidus to temperature T_1, the initial solid formed is of composition C_1^S, coexisting with liquid C_1^L. On further cooling, to T_2, solid C_2^S forms and liquid C_1^L changes composition to C_2^L, in local equilibrium with C_2^S. In the absence of 'back diffusion' the initial solid C_1^S, remains unchanged in composition and the new solid of composition C_2^S grows around the initial solid of composition C_1^S, enveloping it. As cooling proceeds, considering the repetition of these steps, the compositions of the solid formed changes along the solidus and a composition gradient is developed in the solid. In the case shown, where k is < 1, the centres of the crystals e.g. dendrites, will be low in solute, while the solute content of the solid progressively increases. In the system illustrated, which shows a eutectic, the liquid will be enriched in solute up to the eutectic composition, at which stage a eutectic mixture will solidify. In the case of a system containing a peritectic, the solid peritectic product can form as a layer on the primary crystals isolating them from the liquid. The peritectic product, which may be the primary solid solution based on the second component of the system, will then further solidify to form cored crystals, assuming no back diffusion.

The CALPHAD calculation, modelled by computer, uses a series of isothermal steps to model this process and, at very small temperature intervals, is almost completely equivalent to a continuous cooling process. It can take account of values of k that vary during solidification, can be applied to multicomponent systems and can also deal with all the phases that can form in the system under consideration (see, for example in aluminium based alloys Chapter 8). Extensive application of the method just described has shown that the assumption of no 'back diffusion' in the solid holds remarkably well for many material types, such as Al alloy and NiFe-based superalloys. However, there are notable exceptions. For example in low alloy steels, where rapid diffusion of C permits near equilibrium solidification to occur.

8. Selected Case Studies of Ternary Systems

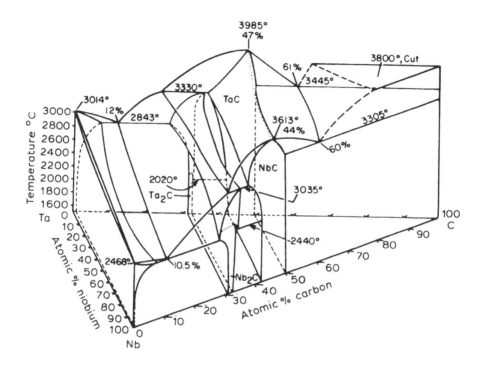

Fig. 8.1 Ternary phase diagram for Nb–Ta–C (Reproduced by permission of Academic Press, Orlando, Florida 32887-6777 from L. E. Toth, *Transition Metal Carbides, Refractory Metal Series 7*, Academic Press, New York, 1971).

8.1 Nb–Ta–C

This system provides an example of complete mutual solid solubility between two refractory metal monocarbides, of interest for their high melting points and hardness. The space model depicted in Figure 8.1[28] shows the Nb–C and Ta–C systems as being of the same type, containing carbide phases MC and M_2C (the latter has two structural modifications). Regions of complete solid solubility link Nb and Ta, Nb_2C and Ta_2C; and NbC and TaC, respectively. The following three-phase reactions occur: L $\rightleftharpoons$ Nb-Ta solid solution + M_2C; L + MC $\rightleftharpoons$ M_2C; L $\rightleftharpoons$ MC + C. The diagram is probably an oversimplification since the literature shows differences in detail of the carbide phase relationships.

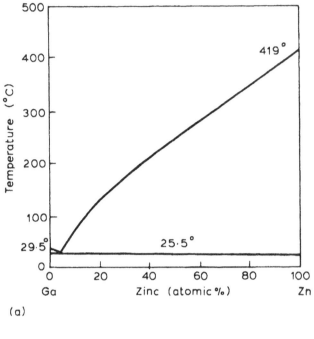

(a)

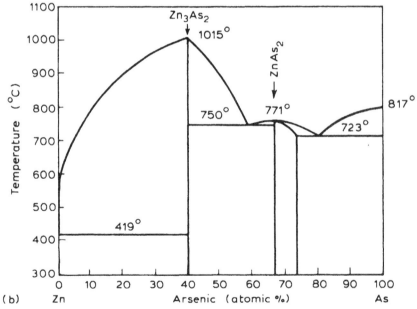

(b)

Fig. 8.2 (a) Ga–Zn binary system, (b) Zn–As binary system, (c) Ga–As binary system and (d) Liquidus surface of As–Ga–Zn system[29, 32] representing the condensed phases in equilibrium with vapour (under closed system conditions). Three congruently melting binary compounds exist: GaAs, Zn_3As_2 and $ZnAs_2$, with associated eutectic reactions: $L \rightleftharpoons Ga + GaAs$ (29.5°C); $L \rightleftharpoons GaAs + As$ (810°C); $L \rightleftharpoons Zn + Zn_3As_2$ (419°C); $L \rightleftharpoons Zn_3As_2 + ZnAs_2$ (750°C); $L \rightleftharpoons ZnAs_2 + As$ (723°C). The Ga–Zn system contains a eutectic reaction: $L \rightleftharpoons Ga + Zn$ (25.5°C).

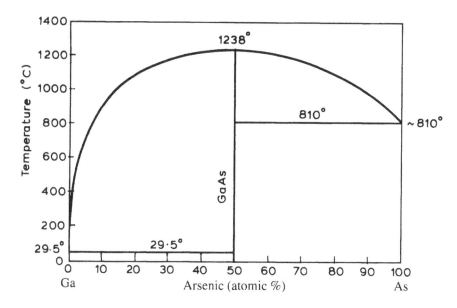

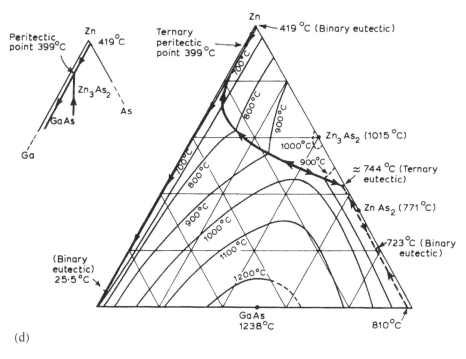

(d)

Fig. 8.2 (Contd.) There are four ternary invariant reactions in each of which the liquid compositions lie very close to binary compositions: L $\rightleftharpoons$ Ga + Zn + GaAs (~20°C); L + Zn$_3$As$_2$ $\rightleftharpoons$ GaAs + Zn (399°C); L $\rightleftharpoons$ GaAs + Zn$_3$As$_2$ + ZnAs$_2$ (~744°C); L $\rightleftharpoons$ GaAs + ZnAs$_2$ + As (720°C); The mutual solid solubilities are very small for all the solid phases in the system, ((d) M. B. Panish, *Journal of Physics and Chemistry of Solids*, 1966, 27, 291, Ga–As–Zn system with permission from Elsevier Science).

Problem

Draw a schematic vertical section from the mid-point of the Nb–Ta system to the C corner of the system and label the phase regions. By reference to this section describe the changes that occur during the melting of an alloy containing 55 at.% Nb, 10 at.% Ta and 35 at.% C.

8.2 Ga–As–Zn

Ternary phase diagrams for systems of Ga-As + an element such as Zn, Sn, Cu, Ag and Au are important in relation to the incorporation of a dopant material into GaAs to produce p-n junctions. Consider the Ga–As–Zn system[29, 30] (Figure 8.2). Elemental Zn, dilute solutions of Zn in Ga or various combinations of Zn and As can be used as source materials in sealed capsules for the diffusion of Zn into the compound. However, alternative source materials have been investigated in the light of phase diagram data.[31, 32] These include compositions lying in a three-phase region (e.g. 5 Ga–50 As–45 Zn (at.%) in the GaAs + Zn_3As_2 + $ZnAs_2$ solid state region at temperatures below ~744°C); within such regions the partial pressures of the components do not vary with composition so that the surface composition and diffusivity depend only on temperature. Such diffusion sources have been found advantageous with respect to obtaining reproducible steep diffusion profiles and planar junction interfaces; also it is not necessary to maintain stringent control of the system composition and capsule volume.

Problems
1. Consider a diffusion source of composition (at.%) 50 Ga–25As–25 Zn placed with a sample of GaAs in an evacuated and sealed silica capsule, and heated to 1000°C. Deduce the composition of the phase in equilibrium with GaAs at this temperature.
2. (a) Draw an isothermal section for the system at 700°C. (b) By reference to this section calculate the proportions by weight of the phases present in an alloy containing 5 Ga–50 As–45 Zn (at.%) equilibrated at 700°C.
3. From the nature of the reactions at 744 and 720°C what can be deduced about the equilibria along the sections GaAs–Zn_3As_2 and GaAs–$ZnAs_2$?

8.3 Au–Pb–Sn

The Au–Pb–Sn system[33] (Figure 8.3), particularly the region between Pb, Sn and the compound $AuSn_2$, is relevant to the use of Pb–Sn solders for joining Au–plated surfaces in the electronics and telecommunications industries. The liquidus has a relatively shallow slope near the Pb-Sn binary system so that substantial solution of Au occurs in molten solder at normal soldering temperatures. The formation of platelets of the brittle compound

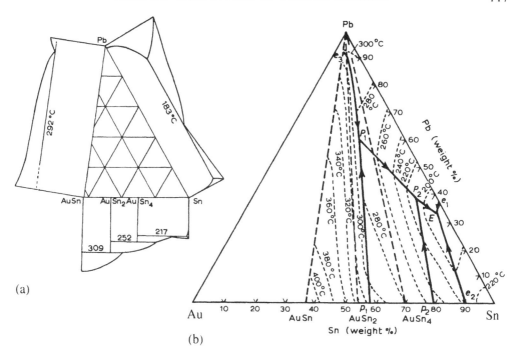

Fig. 8.3 (a) The constituent binary phase diagrams of the AuSn–Pb–Sn ternary system[33] and (b) Liquidus surface of the AuSn–Pb–Sn system involving the solid solutions based on Pb and Sn and the intermetallic compounds AuSn, $AuSn_2$ and $AuSn_4$.[33] The Pb–Sn system contains a eutectic (e_1) (183°C). AuSn is a congruently melting compound (419°C), and the binary system between AuSn and Sn contains the following reactions: $L + AuSn \rightleftharpoons AuSn_2$ (p_1) (309°C); $L + AuSn_2 \rightleftharpoons AuSn_4$ (p_2) (252°C); $L \rightleftharpoons AuSn_4 + Sn$ (e_2) (217°C). The section AuSn–Pb is quasi-binary containing the eutectic $L \rightleftharpoons AuSn + Pb$ (e_3) (292°C). Ternary invariant reactions occur as follows: $L + AuSn \rightleftharpoons Pb + AuSn_2$ (P_1) (275°C); $L + AuSn_2 \rightleftharpoons Pb + AuSn_4$ (208°C); $L \rightleftharpoons Pb + Sn + AuSn_4$ (E) (177°C). Solid solubilities are neglected here. Dashed lines (− − −) represent the sections between Pb and the various Au–Sn compounds[33] (Alan Prince, *Journal of Alloys and Compounds* (Formerly *Journal of Less Common Metals*) 1967, 12, 1078, with permission from Elsevier Science).

$AuSn_4$ occurs on solidification involving the invariant peritectic (P_2) (which does not proceed to completion at the cooling rates encountered in soldering) and the eutectic solidification (e). One procedure for minimizing embrittlement is to keep the thickness of the Au coating below 5 μm.

Problems

1. Describe the solidification under equilibrium conditions of an alloy containing 10 wt.% Au, 40 wt.% Pb, and calculate the proportions of the phases present at 170°C, assuming negligible solid solubility.

2. What percentage of Au can dissolve in a 40% Pb-60% Sn solder at 280°C?

3. Compare the equilibrium solidification behaviour of a 40% Pb, 60% Sn solder used on a gold-plated surface at 220°C with that of a 50% Pb, 50% Sn solder used at 240°C; include reference to the proportions of liquid that undergo the peritectic reaction at P_2 and the proportions of $AuSn_4$ that will form in the two alloys.

8.4 Ni–Al–Cr, Ni–Al–Ti and Ni–Cr–Al–Ti

Nickel-based superalloys are of complex constitution, and contain a range of alloy elements, typically including Al, Co, and refractory metals such as Cr, Ti, Mo, Hf, Ta, and also C. These alloys consist essentially of a nickel-based solid solution (γ), an intermetallic compound based on Ni_3Al (γ'), and carbide phases. The intermetallic sigma (σ) phase may also be present, although alloys are designed to avoid its formation. The ternary system Ni–Al–Cr[34] (Figures 8.4(a-g) forms a useful basis for understanding the relationship between the γ and γ' phases, which are of major importance in the control of mechanical properties: it is also relevant to the role of the β-phase as a constituent of certain surface coatings on superalloy components.

Superalloys typically form primary γ on solidification and commonly contain some γ + γ' eutectic: they are heat-treated to give precipitation of γ' from supersaturated γ. The isothermal sections at 1000 and 750°C (Figures 8.4 (e and f)), together with the vertical section at 75 at.% Ni (Figure 8.4 (g)) illustrate the extensive composition range of γ, and the decrease in solubility of Al in γ with decreasing temperature.[34] Consider, for example, an alloy containing 75 at.% Ni, 10 at.% Al, 15 at.% Cr, solution treatment at, say, 1000°C, followed by rapid cooling, produces supersaturated γ, and ageing at, say, 750°C, produces a fine dispersion of γ' precipitate.

The role of Ti can be taken into account by reference to part of the Ni–Al–Ti system[37-39] (Figures 8.4 (h and i)) which contains a compound Ni_3Ti and to a section through the quaternary Ni-Cr-Al-Ti system at 75 at.% Ni and 750°C (Figure 8.4 (j)).[40] (A quaternary isothermal, isobaric section can be represented by an equilateral tetrahedron, and a section at a constant proportion of one component can be represented by an equilateral triangle.) In Figure 8.4(j) the γ/γ' tie lines lie close to the plane of the section, and the composition point X is close to that of Nimonic 80A(i.e. ~72 Ni, 21 Cr, 2.7 Al and 2.9 Ti (atomic %) + small amounts of other elements). The γ field extends to higher Al and Ti contents with increasing temperature, and at 1000°C Nimonic 80A is well within the γ region for solution treatment. An important trend in wrought and cast superalloy development (particularly in cast alloys) has been to increase the γ' content by increasing the Al + Ti content.

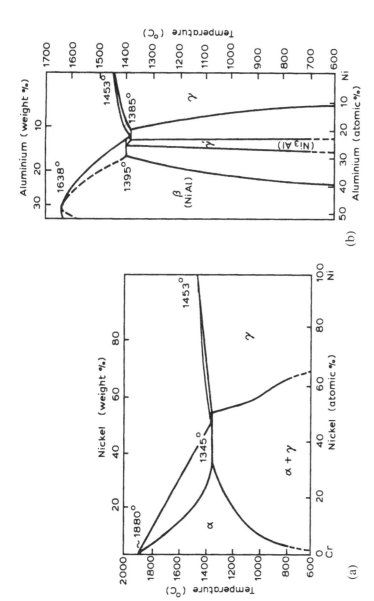

(a)

(b)

Fig. 8.4 (a) Ni–Cr binary system. Note that the α-range data differ from the data used in (c) (e.g. showing a greater solubility at the eutectic temperature) being taken from a more recent source and (b) Part of Ni–Al binary system. (Recent work[35,36] on the binary Ni–Al system has shown that the invariant reactions involving L, γ, γ' and L, β, γ are separated by only a very small temperature interval, of the order of 2°C. It has been suggested that the former reaction is a peritectic: L + γ → γ' instead of a eutectic: L → γ + γ'. The other reaction is now shown as a eutectic: L → β + γ' and not as a peritectic: L + β → γ'. The system also shows a metastable reaction: L → β + γ. It appears that slight differences in alloy composition or solidification conditions e.g. cooling rate, can affect the nucleation process and hence influence which of the reactions occurs. Thus, the Ni–Al phase diagram in Figure 8.4(a) and the Ni–Al–Cr and Ni–Al–Ti liquidus projections in Figures 8.4(c, d and h) would require some modification to match the new information. However, the diagrams from the Second Edition of the book have been retained here).

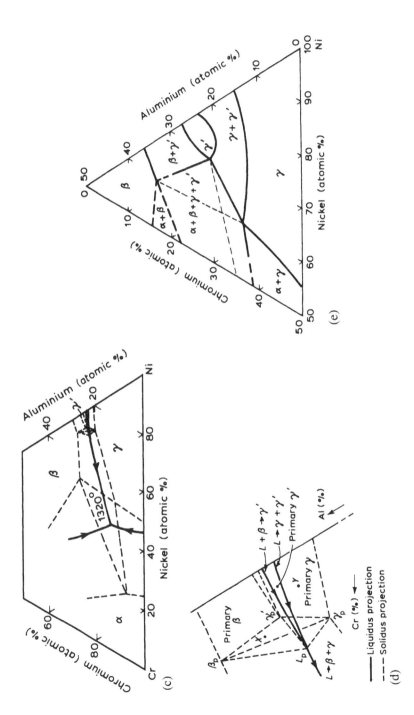

Fig. 8.4 (c) Tentative diagram of fields of primary crystallization for the portion of the Ni–Al–Cr system involving γ (Ni-based solid solution), γ'(Ni₃Al), β (NiAl) and α (Cr-based solid solution). The invariant liquid/solid reactions are: Binary: L + β ⇌ γ' (1395°C); L ⇌ γ + γ' (1385°C); Ternary: L + γ' ⇌ β + γ (1340°C); L ⇌ α + β + γ (1320°C). Also a solid state invariant reaction, β + γ ⇌ α + γ' occurs at ~1000°C (Figures 8.4(e) and A9. The dashed lines represent the solid state constitution including the invariant reaction planes at 1340 and 1320°C. (d) Enlarged view of part of the Ni–Cr–Al system (Figure 8.4(c)) showing the L + γ' ⇌ β + γ reactions and (e) Isothermal section at 1000°C.

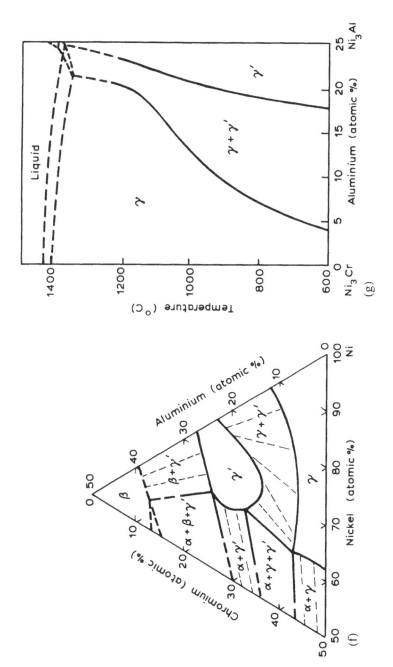

Fig. 8.4 Ni–Al–Cr system (f) Isothermal section at 750°C and (g) Vertical section at 75 at.% Ni.

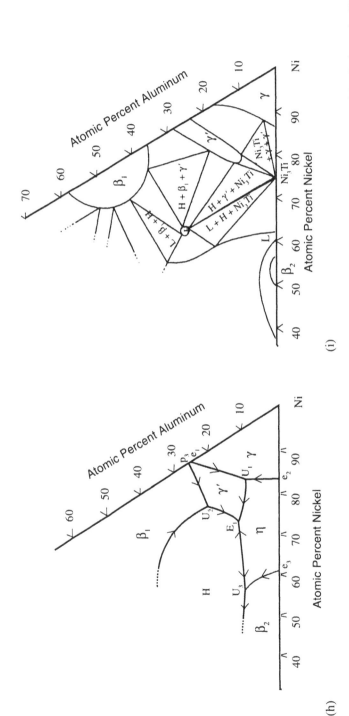

Fig. 8.4 (h) Ni–Al–Ti system. Part of liquidus properties. (The solid phases shown are as follows: γ = Ni-based solid solution, γ' = Ni₃Al (melting incongruently at 1395°C), β₁ = NiAl (melting congruently at ~ 1638°C), β₂ = NiTi (melting congruently at 1310°C), η = Ni₃Ti (melting congruently at 1380°C) and H = Ni₂AlTi (melting congruently) and (i) Ni-Al-Ti system. Part of 1150°C isothermal section. (Reproduced with permission from ASM International, Materials Park, OH 44073-0002, U.S.A). These diagrams are based on the evaluation reported by Lee and Nash.[37] Work reported more recently shows some differences from the paper of Lee and Nash e.g. the invariant reaction E₁ is shown as a peritectic by Zeng et al.[39] An interesting feature reported by Willemin and Durand-Charre[38] is a change from L → γ + γ' eutectic to peritectic: L + γ → γ' with increase in titanium content).

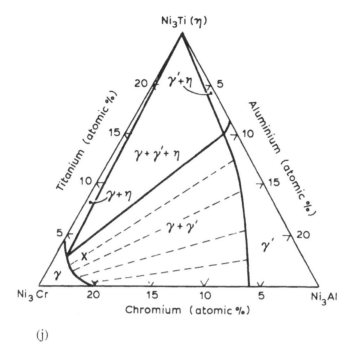

(j)

Fig. 8.4 (j) The Ni$_3$Cr-Ni$_3$Ti-Ni$_3$Al section of the Ni-Cr-Ti-Al system; 750°C isothermal. X corresponds approximately to the composition of Nimonic 80A alloy.

To take account of the influence of other elements on the constitution of superalloys, various ternary and quaternary systems can be consulted. For example, the influence of refractory metals on the γ/γ' equilibrium may be illustrated by systems such as Ni–Al–Mo and Ni–Al–Hf. Sigma-phase formation can be considered in terms of systems such as Ni–Cr–Co and Ni–Cr–Co–Mo, and carbide formation in terms of systems such as Ni–Cr–Mo–C. While such ternary and quaternary phase diagrams provide valuable insights, they do not represent precisely the combined effects of the wide range of elements commercially involved. The PHACOMP procedure for controlling sigma formation and the use of 'polar' phase diagrams have been complementary approaches to the understanding of industrial superalloys.[41] However, more recently CALPHAD techniques have been applied successfully to many multi-component superalloys, including those alloys developed as an important advance for the production of single crystal turbine blades.

Problems

1. By reference to Figure 8.4(d) describe the equilibrium solidification sequences of
   (a) an alloy that initially deposits primary β, and which consists of β + γ + γ' when solidification is completed and (b) an alloy containing 75 at.% Ni, 22 at.% Al and 3 at.% Cr.

2. Draw the three phase regions lying just above and just below the invariant reaction plane at 1000°C to illustrate the transition from the β + γ to the α + γ' equilibrium.
3. Discuss the effect on the solid state equilibria of increasing the Cr content from 0-25 at.% in series of alloys containing (a) a constant Al content of 15 at.%, and (b) a constant Ni content of 75 at.%.
4. Making appropriate reference to the liquidus projection and the 1150°C isothermal section of the Ni–Al–Ti system (Figures 8.4(h and i)), and the other information given in the Figure caption:
   (a) State the nature of the invariant solidification reactions designated in Figure 8.4(h) as p_3, e_1, e_2, U_1, U_2, E_1, e_3, U_3, (b) Describe a possible solidification sequence of a ternary alloy containing (at.%): 30 Al and 10 Ti, cooled under equilibrium conditions from the liquid state to 1500°C.

8.5 Fe–Cr–C, Fe–Cr–Ni and Fe–Cr–Ni–C

The iron-rich corner of the Fe–Cr–Ni–C system forms the basis of the wide range of alloys designated 'stainless steels'. The ternary systems Fe–Cr–C and Fe–Cr–Ni provide a basis for understanding the constitution of these materials, which can be extended by appropriate sections through the quaternary system. Solid state transformations involving the allotropic forms of iron, ferrite (α or δ), bcc, and austenite (γ), fcc, are of major importance. The transition between α-stabilizing (Fe-Cr) and γ-stabilizing (Fe–C and Fe–Ni) phase relationships gives rise to complex regions in the ternary systems.[3, 9a]

The binary systems Fe–Cr, Cr–C, Fe–Ni and Ni–Cr are represented in Figures 8.5(a-c) and 8.4(a). In the Fe–Cr system all alloys solidify as bcc α-phase; the system shows a γ-loop region and beyond 13% Cr γ does not form. Below ~820°C sigma-phase (σ) forms, centred around the approximate equiatomic composition and entering into equilibrium with α over a wide composition range, with an associated embrittling effect. At ~440°C σ decomposes eutectoidally into iron-rich and chromium-rich bcc solid solutions (the chromium-rich solution may be designated α'). The Fe–Ni system contains a peritectic reaction $L + \delta \rightleftharpoons \gamma$. The $\gamma \rightleftharpoons \alpha$ transformation temperature is progressively decreased by Ni additions. The Ni–Cr system shows extensive solid solubility of Cr in Ni (γ-phase) and of Ni in Cr (α-phase) and there is a eutectic reaction $L \rightleftharpoons \alpha + \gamma$.

8.5.1 Fe–Cr–C

The liquidus projection (Figure 8.5(d))[42] shows the following phases: α, γ, graphite, M_3C (based on the Fe–C system), and three carbides ($M_{23}C_6$, M_7C_3, and M_3C_2) based on the Cr–C system and showing extensive solubility for Fe. The solid state constitution is illustrated by isothermal sections at 850°C (Figure 8.5(e) shows the equilibria between the γ-phase and the carbides) and at 700°C (Figure 8.5(f) shows the equilibria between the α-phase and the carbides); equilibria do not occur between M_3C_2, and either γ or α, because of the very high solubility of Fe in M_7C_3. Vertical sections are shown at constant C (0.10%) with variable Cr, and at constant Cr(13%) with variable C[43, 44] (Figures 8.5(g and h)). The solid state constitution includes two invariant reactions:

$$\gamma + M_{23}C_6 \rightleftharpoons \alpha + M_7C_3$$

$$\gamma + M_7C_3 \rightleftharpoons M_3C + \alpha$$

With the addition of C to Fe–Cr alloys the γ-loop is moved to higher Cr levels and the $\alpha + \gamma$ field is widened. The maximum displacement of the phase boundary for fully austenitic alloys is reached at 0.6% C and 18% Cr. Considering the addition of Cr to Fe–C alloys, the γ field is constricted, the eutectoid temperature ($\gamma \rightleftharpoons \alpha + M_3C$) being raised and the C content of the eutectoid being lowered (Figure 8.5(i))[45] *

Problems
1. State the nature of each of the ternary invariant reactions shown in Figure 8.6(a).
2. Describe the equilibrium solidification sequence to just below the solidus of an Fe–13 wt.% Cr–0.1 wt.% C alloy making reference to Figures 8.5(d and g).
3. Samples of an Fe–13 wt.% Cr–0.5 wt.% C alloy are heated for one hour at temperatures of 1200 and 850°C respectively. The samples are then water quenched and tempered for one hour at 700°C. Describe the structural changes that would occur during these treatments, making reference to Figures 8.5(e and h).

8.5.2 Fe–Cr–Ni

This system has been extensively investigated and the data have been critically evaluated.[46] The liquidus and solidus projections for the ternary Fe–Cr–Ni system (Figures 8.5(j and k)) show the transition between the peritectic reaction in the Fe–Ni system and the eutectic in the Ni–Cr system, and there is a temperature minimum not far from the Ni–Cr binary.

The isothermal section at 1000°C (Figure 8.5(l and m)) shows the extensive single phase α and γ regions, together with the two-phase region linking the $\alpha + \gamma$ region bounding the γ-loop in the Fe–Cr system with the $\alpha + \gamma$ region formed by eutectic solidification in the Ni–Cr system. This section forms part of the basis of the Schaeffler diagram for representing the constitution of stainless steels, including the effect of addition of various alloy elements, and used, for example in relation to welding. At temperatures below about 900°C (e.g. at 650°C, Figure 8.5(n)) the $\alpha + \gamma$ region extends into the ternary system from the Fe–Ni and Ni–Cr systems; in the Fe–Ni system the $\alpha + \gamma/\gamma$ boundary

---

* It should be noted that solid state reactions such as eutectoids in ternary systems e.g. $\gamma \rightleftharpoons \alpha + Fe_3C$ involve the presence of a three-phase region (e.g. containing $\gamma + \alpha + Fe_3C$) analogous to the three-phase regions associated with three-phase eutectic regions, such as $L \rightleftharpoons \alpha + \beta$, as discussed in Chapter 4. Concerning the transition between the $\gamma + \alpha$ regions of the γ-loop system (Fe–Cr) to the eutectoid-type system (Fe–C) a detailed discussion is given in Ref. 9a.

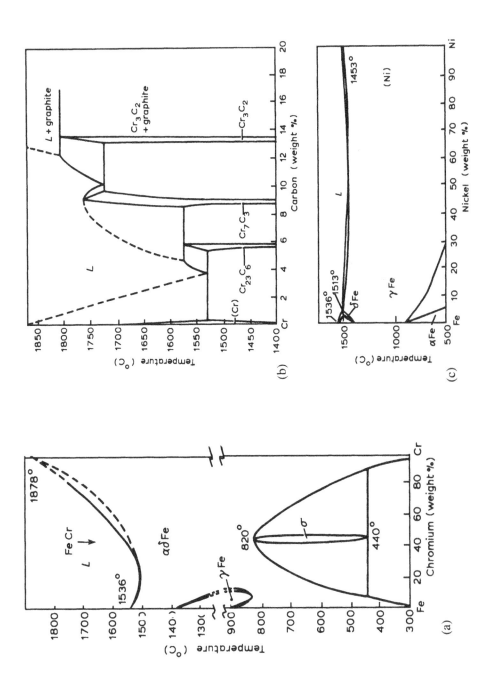

Fig. 8.5 (a) Fe–Cr binary system, (b) Part of Cr–C binary system and (c) Fe-Ni binary system (certain solid state transformation that occur at less than approximately 560°C are not shown here).

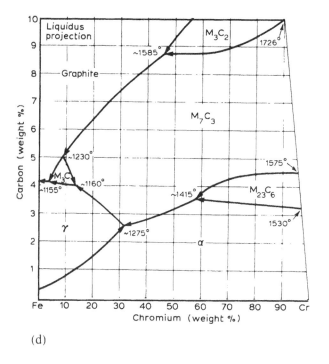

(d)

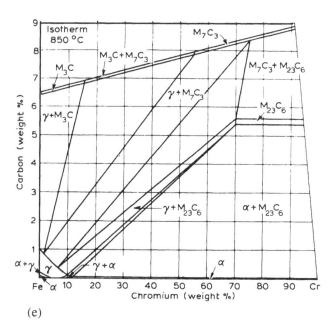

(e)

Fig. 8.5 Fe–Cr–C system[42, 43] (d) Liquidus projection and (e) Isothermal section at 850°C.

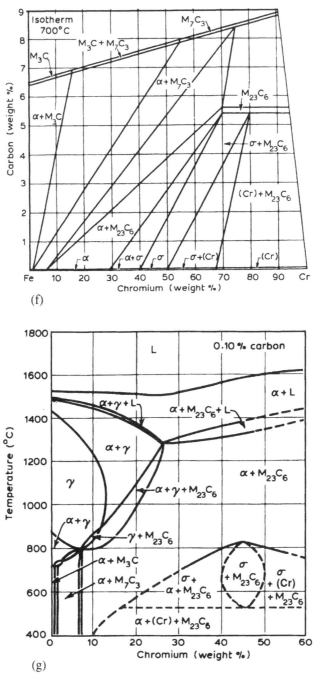

Fig. 8.5 Fe–Cr–C system (f) Isothermal section at 700°C and (g) Vertical section at 0.10% C. The diagram (Ref.42) has been modified here to agree with the form obtained by a CALPHAD calculation in the solid state region to show the invariant reaction at ~510°C involving σ, α, (Cr) and M$_{23}$C$_6$. (d to g.[42] Metals Handbook, 8th Edition. Reproduced with permission from ASM International, Materials Park, OH 44073-0002, USA).

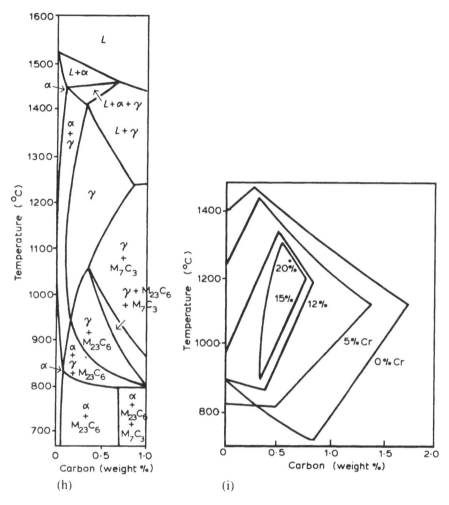

Fig. 8.5 Fe–Cr–C system (h) Vertical section at 13% Cr[43] and (i) Effect of chromium additions on the temperature and carbon content of the $\gamma \rightleftharpoons \alpha + M_3C$ eutectoid.[45] (Note: the data from references 43 and 45 differ in detail concerning the extent of the γ region).

moves to higher nickel content with decrease in temperature. The upper temperature limit for σ formation ~950°C and the equilibria between σ, α (α') and γ at 650°C are shown in Figure 8.5(n).

Among the industrial alloys based on the Fe–Cr–Ni system is the type designated as duplex stainless steels; these are highly formable, corrosion resistant materials, which typically have a composition which allows processing to achieve a structure consisting of a 50/50 mixture of austenite, γ, and ferrite, α. A typical composition is Fe–22Cr–5.5.Ni (wt.%),

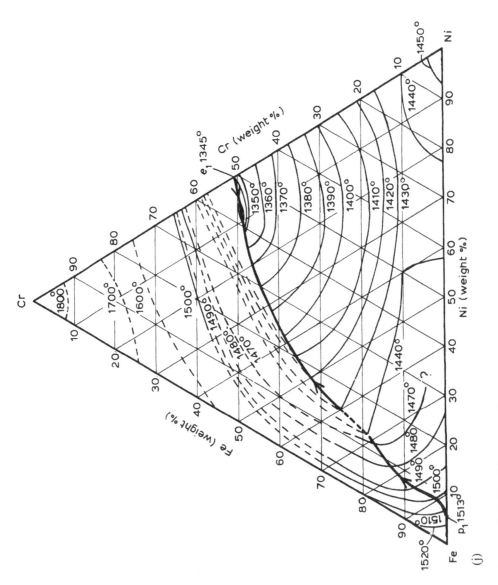

Fig. 8.5 Fe–Cr–Ni system[46] (j) Liquidus projection.

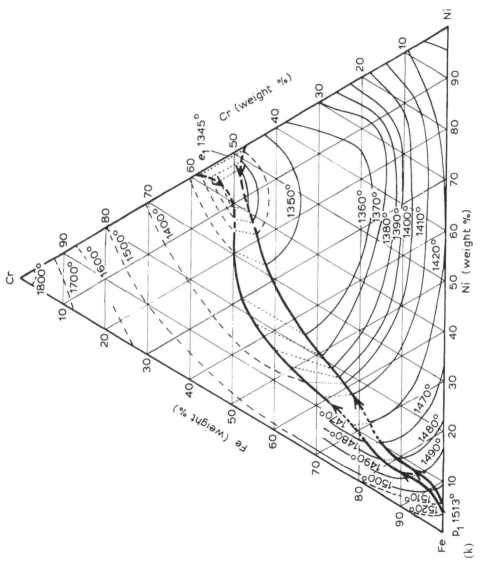

Fig. 8.5 Fe–Cr–Ni system[46] (k) Solidus projection.

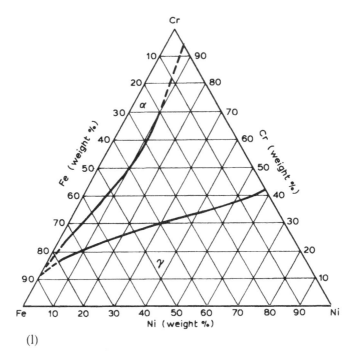

(l)

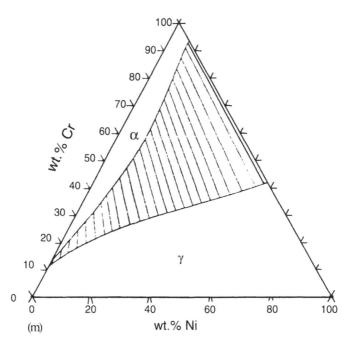

(m)

Fig. 8.5 Fe–Cr–Ni system[46] (l) Isothermal section at 1000°C and (m) Fe–Cr–Ni system. Calculated isothermal system at 1000°C.[13] (Reprinted from N. Saunders and A. P. Miodownik: CALPHAD, Calculation of Phase Diagrams, 1996, p.328 with permission from Elsevier Science).

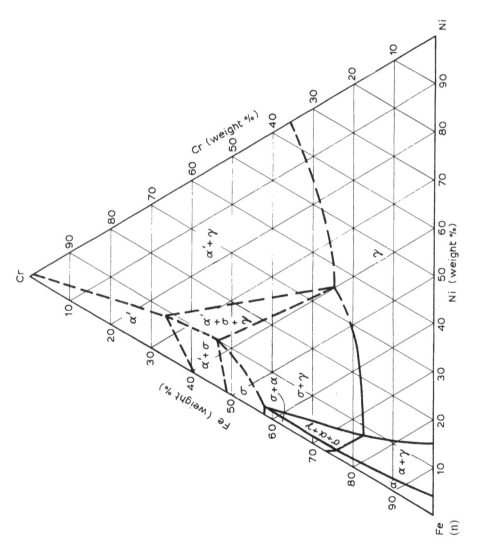

Fig. 8.5 Fe–Cr–Ni system[46] (n) Isothermal section at 650°C (Figures 8.5j, k, l and n Reproduced from Ref. 46).

with relatively small amounts of other elements, e.g.: Mo(3%), Mn(1.7%), Si(0.4%), C(0.024%) and N(0.14%). The constitution of this alloy, including the sequence of solidification and solid state changes can be illustrated by reference to data for the ternary Fe–Cr–Ni system, shown in Figures 8.5(j, k, l, and n) and by data obtained by CALPHAD (Figure 8.5(o)).

From Figures 8.5(j) the experimentally determined liquidus temperature for the ternary alloy is close to 1490°C, Figure 8.5(o) and the primary phase is α. The solidification range is narrow (see Figure 8.5(o)), and during subsequent cooling, the system passes from the α range into the α + γ range (the temperature at which γ starts to form is calculated as 1216°C). The calculated equilibrium proportions of the two phases at 1015°C (e.g. Figure 8.5(o)) are 50/50, which is close to the final annealing temperature of the full composition of the multicomponent alloys mentioned above, and consistent with the experimentally determined 1000°C isothermal section (Figure 8.5(l)). Between 1000 and 730°C, when σ phase formation is calculated to begin (Figure 8.5(o)) the proportion of γ increases at the expense of α. The appearance of σ corresponds to the entering of the α + γ + σ tie-triangle; σ forms at the expense of α.

8.5.3 Fe–Cr–Ni–C

The effect of C on the constitution of an Fe–18Cr–8Ni steel is illustrated by the vertical section in Figure 8.5(p).[47] Rapid quenching from the γ region of the Fe–18Cr–8Ni material with typical carbon levels (e.g. ~0.05-0.1 wt.%) gives a retained austenite structure. Heating within the γ + carbide region can produce carbide precipitation; the preferential precipitation of carbide at the austenite grain boundaries leads to chromium depletion of the adjoining regions with adverse effects on corrosion resistance (e.g. as in 'weld decay').

Problems

1. Draw an isothermal section for the Fe–Cr–Ni system at 1400°C, based on the data in Figures 8.5(j and k) and making necessary assumptions where relevant data are lacking.
2. Describe the equilibrium solidification sequence of an alloy containing 30 wt.% Cr and 15 wt.% Ni, including a calculation of the proportions of the phases present at 1400°C.
3. Describe the changes that occur in an Fe–18 wt.% Cr–12 wt.% Ni alloy (i.e. an 'austenitic' stainless steel) during the solidification and subsequent cooling to 650°C assuming that equilibrium is attained. For this same alloy state what structure would result from rapid cooling from 1000°C. (Note: the M_s is below room temperature).
4. Consider a series of Fe–Cr–Ni alloys containing 5 wt.% Ni and Cr contents ranging from ~20–70 wt.% annealed to equilibrium at 650°C. Describe how the proportion of σ-phase varies with Cr content, making reference to Figure 8.5(n).
5. By reference to the experimentally determined isothermal section at 650°C (Figure 8.5(n)), show that the proportion of σ phase in the Fe–22Cr–5Ni alloy in

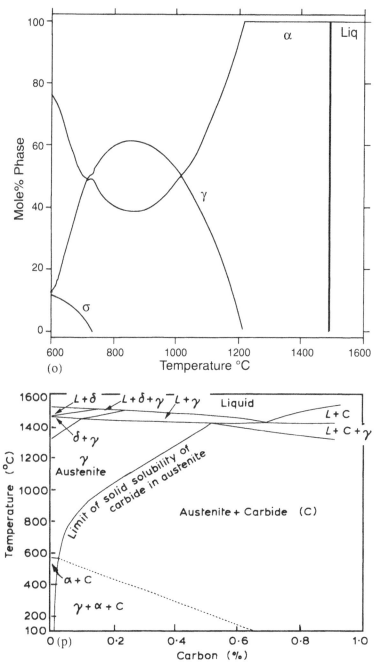

Fig. 8.5 Fe–Cr–Ni system (o) Fe–Cr–Ni system. Calculated mol.% phase vs temperature plots for a Fe–22Cr–5.5 Ni alloy.[13] (Reprinted from N. Saunders and A.P. Miodownik, *CALPHAD, Calculation of Phase Diagrams*, 1963, p328 with permission from Elsevier Science) and (p) Temperature vs composition section showing effect of carbon additions to Fe–18Cr–8Ni steels (C = carbide).[47] Acknowledgments are made to Butterworth Heinemann publishers, a division of Reed Educational and Professional Publisher.

equilibrium at this temperature is consistent with the calculated proportion shown in Figure 8.5(o).

8.6 Fe-O-Si

The situation of pressure acting on a system as a variable is referred to briefly in Chapter 1, and reference may be made to Rhines[6] for the representation of systems taking both temperature and pressure as external variables.

It is important to refer also to the pressure of a system, which is the sum of the partial pressures (p_i) exerted by the gaseous species in the vapour phase generated by incomplete vaporisation of the components (i); these partial pressures are functions of temperature and composition. In many systems the vapour pressure of the condensed phases (liquids and solids) is very low, and, as is the case in most of the systems considered in the book, the vapour phase is ignored.

However, in systems where the elements can exist in more than one oxidation state, the partial pressure of oxygen in the surrounding atmosphere can have a significant influence on the composition of the condensed phases, liquid and solid.[48-50] As a case study to illustrate the principles, consideration is given here to the binary Fe–O and the ternary Fe–O–Si systems, in which the partial pressure of one of the components, namely oxygen, is dominant. The principles are considered here in relation to reactions on cooling from the liquid to the solid state. Some of the oxide compositions within the ternary system are important in the slag chemistry of steel making and other extraction metallurgy processes.

Much detailed work on the Fe–O and Fe–O–Si systems has been carried out including investigations by Darken et al. e.g.[51-52] and Muan[53, 54] and detailed discussion of some of the main features has been presented by Muan[54] and by Ehlers.[10] A more recent review has been presented by Raghavan.[55] The present account is based substantially on that of Ehlers.[10]

In dealing with phase changes occurring during cooling and heating two important situations must be considered. First, 'closed conditions' are such that the total composition of the condensed phases remains constant, viz no oxidation or reduction reaction occurs; then the phase changes in a binary system can be traced on the phase diagram by following vertical lines, as for a binary system for which pressure is assumed constant. Such a situation can be essentially achieved by sealing the condensed phase(s) in an inert, gas-tight container. Alternatively, in the absence of this procedure, there is likely to be sufficient reaction to change the composition of the condensed phase(s).In experimental work, conditions can be applied to control such compositional changes, and the equilibria relationships can be displayed graphically. The oxygen content of the surrounding atmosphere may be controlled by using mixtures of oxygen with an inert gas. Another technique is to keep the condensed phases in contact with either H_2O or CO_2, which decompose on heating to provide oxygen; mixtures of CO and CO_2 or of H_2 and H_2O are often used.

Figure 8.6(a) shows the Fe–O system at a total pressure of 1 atm. indicating the various phase regions. The partial pressure of oxygen is a function of both composition

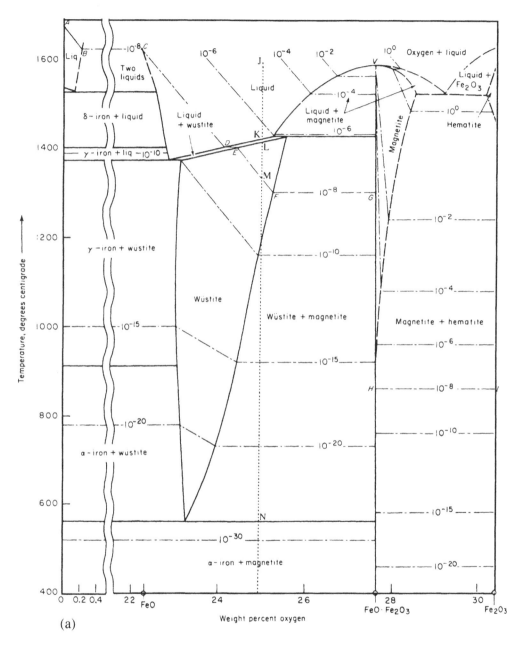

(a)

Fig. 8.6 The FeO binary system. Total pressure 1 atmosphere. Heavy lines represent boundaries of phase regions; light broken lines are O_2 isobars.[10] (After Arnulf Muan, *American Journal of Science*. Reprinted by permission of *American Journal of Science*, reprinted with permission of the American Ceramic Society, P.O. Box 6136, Westerville, OH 43086-6136. © 1960 by the American Ceramic Society. All rights reserved).

and temperature. The two-dimensional diagram shown in Figure 8.6(a) is in fact a projection of the three-dimensional equilibria on to the composition-temperature plane; thus, the diagram is not an isobaric section. The partial pressure of oxygen is plotted on the diagram as isobars, lying parallel to the composition axis, which are shown in the figure as light broken lines, and which coincide with the horizontal tie-lines in a two-phase region. When two condensed phases coexist with a gas phase, there is only one degree of freedom; consequently choice of temperature will fix the composition of the condensed phases and also the oxygen partial pressure. In an area of the diagram (Figure 8.6(a)) where only one condensed phase exists together with the gas phase, there are two degrees of freedom, so that the oxygen pressure and the temperature can vary simultaneously; the oxygen isobars cross such an area 'diagonally' instead of horizontally. The Fe–O diagram determined at 1 atm. pressure shown in Figure 8.6(a) should not strictly be called an equilibrium diagram, but, for practical purposes it may coincide with a diagram determined taking account of partial pressure; however, there can be significant differences in areas of the diagram where the partial pressure is appreciable, e.g. for the high temperature equilibria between magnetite and haematite.

Consider now the process of solidification and solid state cooling under conditions of constant partial pressure of oxygen of 10^{-8} atm., following the relevant isobar, with the bulk composition of the condensed phases changing by gain or loss of oxygen as reactions with the gas phase occur. Point A in the figure represents the highest temperature for this particular value of oxygen partial pressure, and the only composition that can exist at A at this partial pressure is pure iron. As heat is removed from the system, during cooling the liquid iron(A) takes some oxygen into solution and changes its composition to B. With continued removal of heat, oxygen solution proceeds isothermally; a second liquid(C) forms until all of the liquid is of composition C, essentially FeO. Between the points B and C the relative proportions of the two liquids depend on the heat content of the system and cannot be determined from this phase diagram. As the heat removal continues, the liquid becomes richer in oxygen, and attains the composition indicated by point D, when at ~1400°C, solid wüstite (composition E) begins to form, and on cooling this phase is enriched in oxygen to point F. Then, with further removal of heat, magnetite begins to form at the expense of wüstite, until at G no wüstite remains; cooling then brings the magnetite composition to H (near to 850°C). At H the solid undergoes the change from magnetite to haematite, (composition I) which then cools to room temperature.

Consider next, instead of constant partial pressure of oxygen, the case of a closed system, in which oxygen is neither removed or added. For example, assume a bulk composition, point J in the diagram, corresponding to ~25 wt.% oxygen cooled from the liquid state; solidification follows the relevant vertical line in Figure 8.6(a); to maintain a fixed total composition of condensed phases the partial pressure of oxygen decreases with temperature. During cooling, the formation of wüstite begins at the liquidus, point K; the first solid is slightly enriched in oxygen, but as cooling proceeds, the solid wüstite reaches the original solid composition at the solidus, L. At this temperature the oxygen partial pressure is somewhat less than 10^{-6} atm. On further cooling, the partial pressure progressively decreases; magnetite starts to form at M and then, at N, when the wüstite has been depleted in oxygen to the eutectoid composition, the reaction wüstite → α-iron + magnetite takes place.

The same principles as described above for the binary system apply to ternary compositions from the $FeO-Fe_2O_3-SiO_2$ system, but with the added complexity that stems from the extra degree of freedom arising from the third component of the system. Consequently, temperatures may change at constant oxygen pressure across the liquidus surface of the ternary system; oxygen isobars appear as univariant curves across the surface, (shown dashed on the liquidus surface in Figure 8.6(b)) isotherms are also shown as full lines. Figure 8.6(c) shows the solid state compatibility regions on completion of solidification.

Consider first the solidification of an alloy of composition A (Figure 8.6(d)) with constant total composition of the condensed phases; in the solid state this alloy consists of a mixture of fayalite + wüstite; i.e. the composition lies in the fayalite + wüstite region of Figure 8.6(c). Solidification commences when the liquidus temperature is reached; the composition of the wüstite that begins to form can be indicated by a point such as D. Then, as shown schematically the liquid composition changes along a curved path on the liquidus surface as the temperature decreases, (according to the principle discussed in Chapter 3); the wüstite composition also changes, and at any given temperature during this stage of solidification, there will be a tie-line joining the liquid and wüstite compositions and passing through point A. As solidification proceeds the liquid composition reaches point E, which lies on the univariant curve representing the eutectic reaction: $L \rightarrow$ wüstite + fayalite. This stage corresponds to the tie line EAF, where point F represents the wüstite composition. The next stage of solidification is the occurrence of the $L \rightarrow$ fayalite + wüstite eutectic reaction, the progress of which as the temperature decreases can be considered in relation to the L + fayalite + wüstite tie-triangle. At the stage where the eutectic reaction is about to begin, the tie-line EAF forms one side of the tie-triangle. However, as the reaction proceeds, the fayalite composition remains constant, while the liquid composition moves along the eutectic value to reach point C. Also the wüstite composition changes in the direction of higher oxygen content to eventually reach point B. Thus, with decrease in temperature, the oxide alloy composition, A, comes to lie inside the tie-triangle until the eutectic reaction is completed when the alloy composition, A, lies on the tie-line representing the fayalite-wüstite(B) side of the tie-triangle (Figure 8.6(d)). The solidification process is completed before the liquid composition reaches the ternary invariant point representing the eutectic: $L \rightarrow$ wüstite + fayalite + magnetite. As in the case of the binary Fe-O system, for solidification under constant total composition, the partial pressure of oxygen will change during cooling to correspond to the relevant isobars.

The solidification of the same composition, A, is now considered for conditions of fixed partial pressure of oxygen, in which the bulk composition of the condensed phases changes by adding or releasing oxygen to the surroundings. To explain the changes in bulk composition, reference is made to the $Fe-Si-O_2$ system, of which the $FeO-Fe_2O_3-SiO_2$ system is a part (Figure 8.6(e)). The change in composition of any condensed phase can occur either towards or away from the oxygen corner of the system (Figure 8.6(e)).

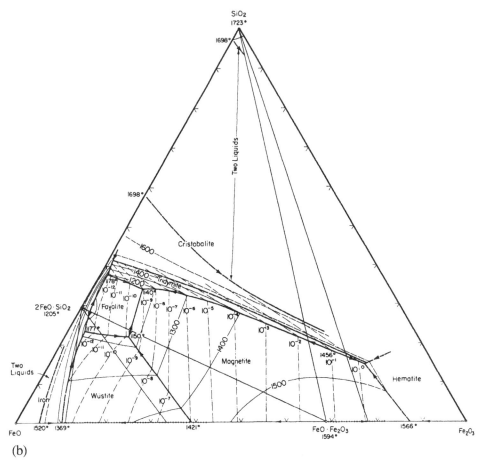

(b)

Fig. 8.6 (b) The FeO–Fe$_2$O$_3$–SiO$_2$ system; wt.%. The heavy lines represent reaction curves on the liquidus projection; the light lines represent isothermal curves and the light broken lines represent equal partial pressures of oxygen(atm.) at the liquidus.[10, 56] (Reprinted by permission of *American Journal of Science*, reprinted with permission of the American Ceramic Society, P.O. Box 6136, Westerville, OH 43086-6136. © 1960 by the American Ceramic Society. All rights reserved). The phases derived from the binary FeO–Fe$_2$O$_3$ system for which primary regions of solidification are shown on the ternary liquidus are: iron, wüstite, magnetite and haematite. In the higher silica content region of the ternary system primary regions of tridymite and cristobalite are shown and there is a region of liquid immiscibility. Another feature of the system, important in the present discussion, is the compound fayalite, 2FeO · SiO$_2$ for which a region of primary solidification is shown on the liquidus. Figure 8.6(b) shows that this compound is located on the FeO–SiO$_2$ axis, but it should be born in mind that the FeO–SiO$_2$–Fe$_2$O$_3$ section forms part of the complete Fe–O–Si system. Thus, the equilibria in which fayalite participates extend beyond the FeO–SiO$_2$–Fe$_2$O$_3$ section into the Fe–O–SiO$_2$ part of the Fe–O–Si system. The various investigations reported on equilibria involving fayalite (e.g. Refs. 57 and 58) show that, in contact with iron the compound contains ~2 wt.% Fe$_2$O$_3$, and melts incongruently. The equilibria involving fayalite and iron, are not demonstrated fully in Figure 8.6(b), although the steep nature of the iron-rich liquidus is shown and the region of two liquids is indicated. The thermodynamically calculated isothermal sections shown in Figures 8.6(h-n) provide information beyond the

e.g. starting with a bulk composition A, the change can occur along the line drawn from the oxygen corner through the bulk composition, either towards B or C, while an initial composition, D, might change towards E or F. It is important to note (Chapter 1. Figure 5) that the ratio of Fe/Si remains constant along these lines. Figure 8.6(f) shows an equilateral triangle to represent FeO, Fe_2O_3 and SiO_2 as the components. An example of a straight line within this triangle which represents a fixed Fe/Si ratio and a variable oxygen content is obtained by linking the compositions $2FeO \cdot SiO_2$ and $Fe_2O_3 \cdot SiO_2$, which each correspond to Fe/Si contents of 2/1 in mol.%. This line is also shown plotted in wt.%. It should be noted that in this case, the wt.% line is essentially parallel to the base of the triangle; thus with a wt.% plot of the FeO-$Fe_2O_3 \cdot SiO_2$ system, the bulk phase composition changes along a line parallel to the base of the triangle, as reactions occur by which the condensed phases give up or take up oxygen from the surroundings. Thus the line from $2FeO \cdot SiO_2$ to $Fe_2O_3 \cdot SiO_2$, plotted in wt.%, represents the path on which the bulk alloy composition changes during solidification.

With reference to Figure 8.6(g), consider alloy composition A, which is located in the primary wüstite region of the liquidus and on the P_{O_2} line of 10^{-9} atm. During cooling from the liquid state, wüstite begins to form when the liquidus temperature is reached and the wüstite would initially have a composition such as that of point C. As cooling

---

Fig. 8.6 (b) Caption continued.
boundaries of the FeO–SiO_2–Fe_2O_3 section. Also the detailed review by Raghavan[55] includes information on regions beyond that shown in Figure 8.6(b).

Figure 8.6(b) shows four-phase invariant reactions during solidification as follows:

$L + \gamma$ - Fe	$\rightarrow$ fayalite	+ wüstite	1177°C
$L + \gamma$ - Fe	$\rightarrow$ fayalite	+ silica(tridymite)	1178°C
$L \rightarrow$ fayalite	+ wüstite	+ magnetite	1150°C
$L \rightarrow$ fayalite	+ tridymite	+ magnetite	1140°C
$L \rightarrow$ magnetite	+ tridymite	+ haematite	

Note the saddle-points on the magnetite–fayalite section at 1150°C corresponding to the eutectic formation of fayalite and magnetite and at a higher, but unspecified temperature, associated with the incongruent melting of fayalite, which involves γ-iron. The arrows on the liquidus isotherm also indicate a saddle-point on the tridymite–magnetite eutectic curve near to the silica-magnetite-haematite invariant eutectic.

The review by Raghavan[55] shows some differences for the four phase invariant reaction temperatures; also the reaction shown above as: $L \rightarrow$ fayalite + wüstite + magnetite is presented as a peritectic: $L +$ wüstite $\rightarrow$ fayalite + magnetite.

The difficulties associated with the experimental determination of the complex phase diagram should be born in mind, particularly noting the relatively narrow temperature range within which the first four of the invariant reactions listed occur.

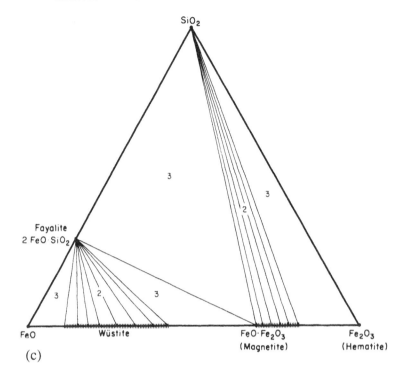

(c)

Fig. 8.6 (c) FeO–Fe$_2$O$_3$–SiO$_2$ system. wt.%. Solid state compatibility fields at the completion of solidification.[10] (Reprinted by permission of *American Journal of Science*, reprinted with permission of the American Ceramic Society, P.O. Box 6136, Westerville, OH 43086-6136. © 1960 by the American Ceramic Society. All rights reserved).

proceeds, with the partial pressure maintained constant, the liquid composition must remain throughout the whole solidification process on the 10^{-9} atm. partial pressure isobar, and moves over the liquidus to reach point *B*, when the wüstite composition will have changed to reach a point such as *D*. The overall bulk composition of the two condensed phases moves along the line *XX'* as oxygen is acquired to reach point *E*, which lies on the tie-line *BD*; the lever rule gives *BE* and *ED* as representing the proportions of wüstite, *D*, and liquid, *B*. When the liquid composition reaches *B*, it must cross into the magnetite primary field, and the wüstite already formed changes to magnetite. This change occurs at a fixed temperature and partial pressure of oxygen, and the bulk composition of the condensed phases changes from *E* to *G* (this point *G* lies on *XX'* and on the line joining *B* and the composition of magnetite. As cooling proceeds at a fixed temperature, the liquid composition follows the 10^{-9} atm. isobar across the magnetite field with continuous formation of magnetite, until the liquid reaches composition *F* on the magnetite-tridymite boundary curve, when the condensed phase composition is *H*; magnetite and tridymite

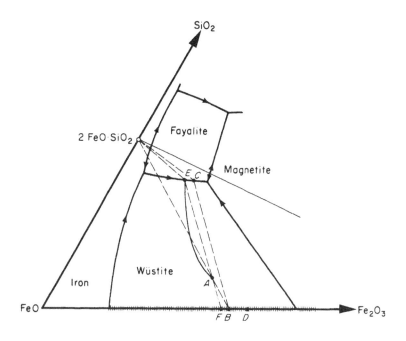

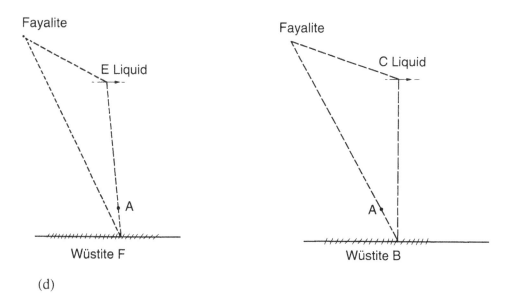

(d)

Fig. 8.6 (d) FeO–Fe$_2$O$_3$–SiO$_2$ system. Liquidus projection showing schematically a solidification path for an alloy with a composition in the primary wüstite field.[10] Alloy A. Tie-triangles showing compositions of phases at the start and end of the $L \rightarrow$ fayalite + wüstite reaction. (Arnulf Muan, *American Journal of Science*, Reprinted by permission of *American Journal of Science*).

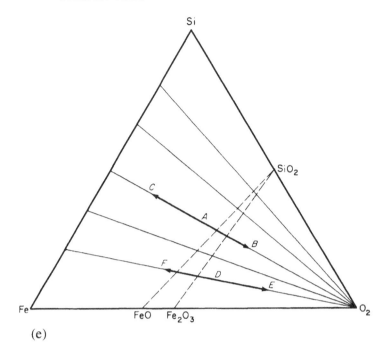

(e)

Fig. 8.6 (e) Fe–Si–O$_2$ system showing how change of oxygen content leads to change of alloy composition towards or away from the oxygen corner of the composition triangle. Bulk composition A with oxygen enrichment moves towards B and with oxygen depletion towards C; correspondingly bulk composition D moves towards E and F respectively.[10] (From The Interpretation of Geological Phase Diagrams by Ernest G. Ehlers copyright, 1972. W. H. Freeman and Company. Used with permission).

form together until the liquid is consumed. It should be noted that F is the lowest temperature on the 10^{-9} atm. isobar, and therefore liquid cannot exist below this point. As magnetite and tridymite form, the composition of the condensed phases changes from H to I. Thus, as solidification of alloy has proceeded, the condensed phase composition has changed to I, which lies on the line joining the composition of magnetite to that of SiO$_2$.

Figures 8.6(h-l) show thermodynamically calculated isothermal sections of the FeO–Fe$_2$O$_3$–SiO$_2$ system (wt. fraction) at temperatures of 1330, 1200, 1190, 1170 and 1160°C respectively, and an isothermal section at 1190°C in mole fraction, (all are at 1.01325^5 Pa(1 atm.) pressure.[*] Three-phase regions are shown shaded, while two-phase

[*] The isothermal sections (Figs. 8.6 h-n) were calculated by Professor B.B. Argent, University of Sheffield, and Dr A. T. Dinsdale, National Physical Laboratory (NPL), Teddington, U.K., using, MTDATA. The data, including the cover illustration, are used with acknowledgements to Dr. A.T.Dinsdale and Dr.J.Gisby, NPL.

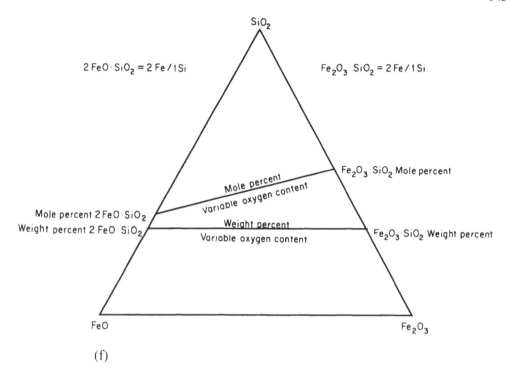

(f)

Fig. 8.6 (f) FeO–Fe₂O₃–SiO₂ composition triangle. Line joining 2FeO · SiO₂ to Fe₂O₃ · SiO₂ (mol.%) represents compositions having the same Fe/Si ratio, but differing oxygen content, increasing on moving towards Fe₂O₃ · SiO₂. The situation is also shown using a wt.% plot; the line joining 2FeO · SiO₂ to Fe₂O₃ · SiO₂ in this case lies essentially parallel to the FeO–Fe₂O₃ side of the composition triangle, so that differences in oxygen content of an alloy can be followed along this line in wt.%.[10] (From *The Interpretation of Geological Phase Diagrams* by Ernest G. Ehlers copyright, 1972. W.H. Freeman and Company. Used with permission.)

regions are shown unshaded, with selected tie-lines. These sections illustrate some details of the liquid-solid phase relationships which have not been shown in the diagrams discussed above. They illustrate various stages of solidification of the liquid oxide alloys and when used with actual calculated values of the proportions of phases at various temperatures, (see for example as shown in Problem 2 below), precise values can be deduced of the temperatures for the start and finish of various reactions such as liquidus and solidus.

In the literature for the FeO–Fe₂O₃–SiO₂ system and many other ceramic systems, the phase diagrams have traditionally been commonly presented in wt. fractions or % s (see e.g. publications of the American Ceramic Society, such as Ref. 7) and this practice is followed here. However, there is an advantage in presenting compositions in terms of

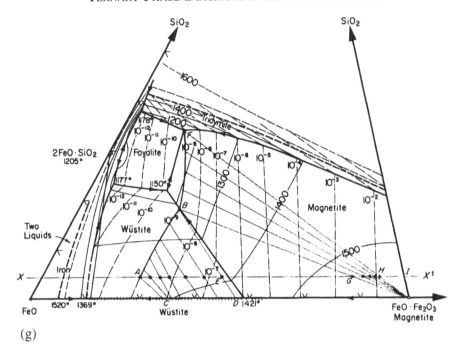

(g)

Fig. 8.6 (g) FeO–Fe$_2$O$_3$–SiO$_2$ system. Features of liquidus to illustrate the solidification of alloy *A* (primary wüstite) and showing change in oxygen along line XX^1 together with liquidus isothermals and curves of constant partial pressure of oxygen.[10] (Reprinted by permission of *American Journal of Science*, reprinted with permission of the American Ceramic Society, P.O. Box 6136, Westerville, OH 43086-6136. © 1960 by the American Ceramic Society. All rights reserved).

moles to display stoichiometric relations involving compounds. An illustrative example is included here at 1190°C (Figure 8.6(k)) and comparison with the wt. fraction diagram at the same temperature (Figure 8.6(j)) show the slopes of the tie-lines differ in direction in the wüstite-fayalite region.

With currently available computer software, such as that used here, plots can readily be made in either wt. or mole fractions from the thermodynamic data; furthermore other important features such as liquidus projections can be plotted.

It is also of interest to note that it is now common to refer to phases for this system using a terminology which denotes the 'generic' structure types. Thus, wüstite would be referred to as halite, magnetite as spinel, haematite as corundum, fayalite as olivine and the liquid as liquid oxide.

To show the FeO–SiO$_2$–Fe$_2$O$_3$ system in the context of the Fe-O-Si system Figure 8.6(n) presents the thermodynamically calculated isothermal section at 1100°C for the region of the Fe–Si–O system (at 1.01325 × 10^5 Pa (1 atm.) pressure. The Fe-rich

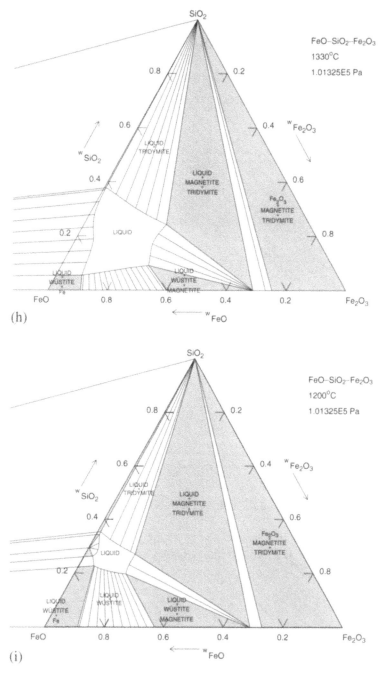

Fig. 8.6 (h) FeO–Fe₂O₃–SiO₂ system. Thermodynamically calculated isothermal section (see footnote) at 1330°C and 1.01325 x 10⁵ Pa(1 atm.) plotted in wt. fraction. Acknowledgements for 8.6 h-n to Prof. B.B.Argent, Dr.A.T. Dinsdale and Dr.J.Gisby and (i) As in Figure 8.6(h), but at 1200°C. (Note: comparison of this isothermal section with that experimentally determined by Muan and Osborn[49] for the same temperature shows good agreement).

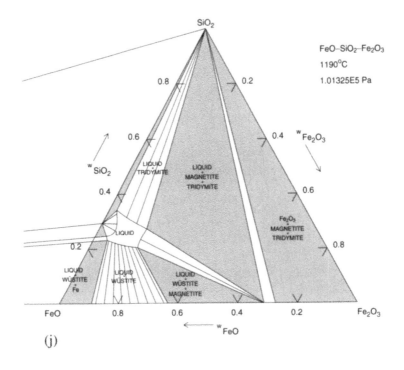

(j)

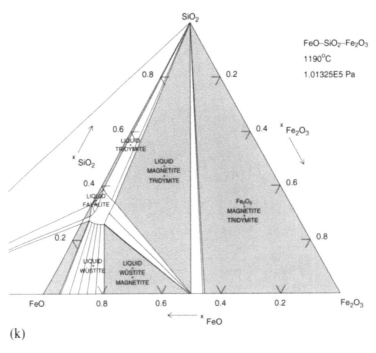

(k)

Fig. 8.6 (j) As in Figure 8.6(i), but at 1190°C and (k) As in Figure 8.6(j), but in mole fraction.

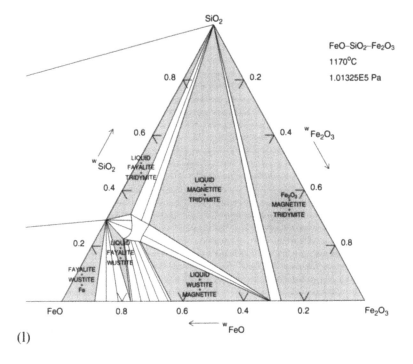

(l)

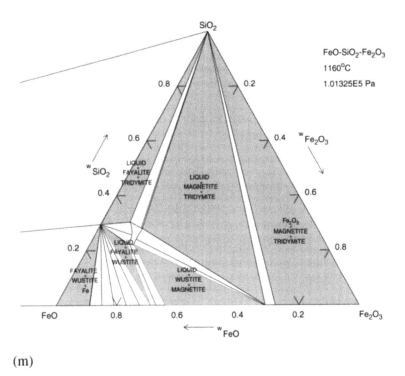

(m)

Fig. 8.6 (l) As in Figure 8.6(h), but at 1170°C and (m) As in Figure 8.6(h), but at 1160°C.

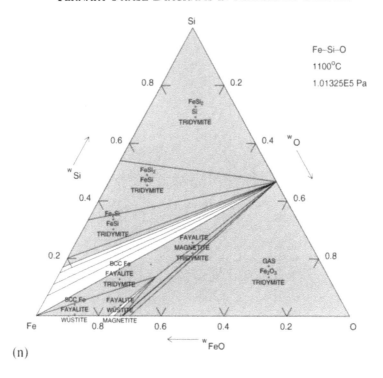

(n)

Fig. 8.6 (n) Isothermal section of the Fe-O-Si system at 1100°C at 1·01325 x 10⁵ Pa(1 atm.) plotted in wt. fraction including the equilibria involving Fe-Si compounds. The section consists predominantly of three-phase regions involving solid phases.

region of the system involving equilibria with Fe is shown, as is also the existence of the gaseous phase in the oxygen rich region.

Problems

1. Consider a slag from the $FeO-Fe_2O_3-SiO_2$ system whose composition lies at the point of intersection of the fayalite-magnetite line with the 1500°C isothermal on the liquidus projection shown in Figure 8.6(b). Assuming that 1 tonne of this slag in the liquid state at 1550°C is slowly cooled, under 'closed equilibrium' conditions, to 1400°C, determine the mass of solid magnetite that will be present at this temperature.

2. Consider an alloy containing (wt. fraction) 0.720 FeO, 0.080 SiO_2, 0.200 Fe_2O_3 (which on completion of solidification under closed conditions consists of wüstite + fayalite); deduce the temperatures and phase compositions (wt. fraction) corresponding to the points D, F, E, B and C shown schematically for alloy A in Figure 8.6(d). Make appropriate reference to the thermodynamically calculated isothermal sections for the $FeO-Fe_2O_3-SiO_2$ system (Figures h, i, j, l and m) (all at 1.01325×10^5 Pa(1 atm.) pressure, together with the data shown in the Table opposite.

Phases Present at Various Temperatures, and at 1.01325×10^5 Pa (1 atm) Pressure, in the range 1350-1150°C for an alloy containing wt. fraction 0.720 FeO, 0.080 SiO$_2$, 0.200 Fe$_2$O$_3$ (Calculated Thermodynamically). Acknowledgements to Dr. A.T. Dinsdale, NPL, for providing the data.

T°C	Liquid FeO			Wüstite			Fayalite		
	FeO	SiO$_2$	Fe$_2$O$_3$	FeO	SiO$_2$	Fe$_2$O$_3$	FeO	SiO$_2$	Fe$_2$O$_3$
1350	0.7200	0.0800	0.2000						
1330	0.7200	0.0800	0.2000						
1327	0.7200	0.0800	0.2000						
1326.5	0.7200	0.0800	0.2000						
1326.3	0.7200	0.0801	0.1999	0.7286	0.0000	0.2714			
1326	0.7200	0.0804	0.1996	0.7288	0.0000	0.2712			
1320	0.7189	0.0871	0.1940	0.7325	0.0000	0.2675			
1310	0.7160	0.0985	0.1855	0.7324	0.0000	0.2626			
1300	0.7120	0.1103	0.1777	0.7411	0.0000	0.2589			
1290	0.7074	0.1223	0.1702	0.7437	0.0000	0.2563			
1280	0.7026	0.1345	0.1629	0.7455	0.0000	0.2545			
1270	0.6979	0.1465	0.1556	0.7466	0.0000	0.2534			
1260	0.6935	0.1581	0.1484	0.7471	0.0000	0.2529			
1250	0.6898	0.1688	0.1414	0.7472	0.0000	0.2528			
1240	0.6866	0.1785	0.1348	0.7471	0.0000	0.2529			
1230	0.6841	0.1872	0.1287	0.7468	0.0000	0.2532			
1220	0.6820	0.1949	0.1232	0.7465	0.0000	0.2535			
1210	0.6802	0.2017	0.1181	0.7462	0.0000	0.2538			
1200	0.6786	0.2079	0.1135	0.7459	0.0000	0.2541			
1190	0.6772	0.2136	0.1092	0.7456	0.0000	0.2544			
1180	0.6758	0.2188	0.1054	0.7455	0.0000	0.2545			
1170	0.6745	0.2237	0.1018	0.7453	0.0000	0.2547			
1168	0.6743	0.2246	0.1011	0.7453	0.0000	0.2547			
1167.5	0.6742	0.2248	0.1010	0.7453	0.0000	0.2547			
1167.2	0.6742	0.2250	0.1009	0.7453	0.0000	0.2547			
1167	0.6739	0.2250	0.1010	0.7447	0.0000	0.2553	0.7051	0.2949	0.0000
1166.5	0.6724	0.2249	0.1027	0.7403	0.0000	0.2597	0.7051	0.2949	0.0000
1166	0.6709	0.2247	0.1044	0.7360	0.0000	0.2640	0.7051	0.2949	0.0000
1165.5	0.6694	0.2246	0.1060	0.7317	0.0000	0.2683	0.7051	0.2949	0.0000
1165	0.6679	0.2245	0.1076	0.7275	0.0000	0.2725	0.7051	0.2949	0.0000
1164.8	0.6673	0.2245	0.1083	0.7258	0.0000	0.2742	0.7051	0.2949	0.0000
1164.5				0.7255	0.0000	0.2745	0.7051	0.2949	0.0000
1164				0.7255	0.0000	0.2745	0.7051	0.2949	0.0000
1160				0.7255	0.0000	0.2745	0.7051	0.2949	0.0000
1150				0.7255	0.0000	0.2745	0.7051	0.2949	0.0000

8.7 Fe–O–S

The system (Figure 8.7(a-d) is of interest in relation to the formation of oxysulphide inclusions in steels.[59] Consider, for example, iron-rich alloys whose compositions (expressed in terms of oxygen–sulphur ratio) are higher than the line drawn from Fe through point C (Figure 8.7(c)). In such alloys, e.g. composition X, the solidification process involves the liquid miscibility gap. For example, alloy X (say 0.1 wt.% O, 0.2 wt.% S) will commence solidification by the formation of δ-iron crystals. The liquid composition changes along the extension of the line joining Fe and X, i.e. it becomes enriched in oxygen and sulphur until its composition comes to lie on the monotectic curve at point Y. The monotectic reaction $L_1 \rightarrow$ Fe $+ L_2$ commences as cooling proceeds (designated L_1 for liquid lying on the portion AC of the monotectic curve and L_2 for liquid lying on the BC portion of this curve). As the reaction proceeds the compositions of L_1 and L_2 move along the curves AC and BC, respectively. Under conditions of true equilibrium, L_1 is consumed; then L_2, on cooling, deposits iron while its composition changes on the extension of the line joining Fe and X. The composition of L_2 meets the eutectic valley $L \rightarrow$ Fe $+$ FeO; this eutectic reaction proceeds until the liquid composition reaches the ternary eutectic point E, when the reaction $L \rightarrow$ Fe $+$ FeO $+$ FeS completes the solidification.

In typical conditions of non-equilibrium solidification involving dendritic growth, the solidification process is modified significantly (Figure 8.7(d)). When the formation of L_2 occurs, regions of this liquid become isolated by the growing dendrites, and no mass transport occurs between these regions. As solidification proceeds the composition of L_1 changes along curve AC while a series of droplets of L_2 form slightly richer in sulphur and of decreased oxygen/sulphur ratio, compared with the first formed droplets of L_2. In general, the compositions of L_1 and L_2 will move to point C, so that the compositions of the series of liquid droplets span the range up to C. With continued cooling each droplet of L_2 solidifies as an isolated region by further separation of iron. The composition of the liquid in each droplet changes on a path extending radially from the Fe corner of the system. These individual paths will intersect either the $L \rightarrow$ Fe $+$ FeO eutectic valley or the Fe $+$ FeS eutectic valley (or for one particular composition will intersect the ternary eutectic point). Solidification of each droplet is completed by the Fe $+$ FeO $+$ FeS ternary eutectic reaction. Concerning liquid L_1, when its composition has reached point C, its solidification proceeds by the separation of iron, followed by the reactions $L \rightarrow$ Fe $+$ FeS and $L \rightarrow$ Fe $+$ FeO $+$ FeS. Thus, under these conditions of non-equilibrium solidification, the oxysulphide inclusions show a range of compositions and hence structural features which is in contrast to the uniform composition of inclusions that would form under equilibrium conditions.

Problems
1. By reference to schematic diagrams of tie-triangles, show how L_1 is consumed during the monotectic reaction in alloy X, assuming equilibrium conditions.

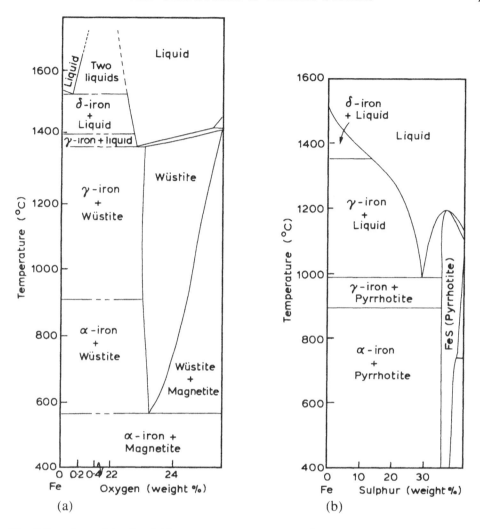

Fig. 8.7 (a) Part of the Fe–O binary system and (b) Part of the Fe–S binary system.

2. For alloys undergoing the monotectic reaction, assuming solidification conditions involving isolation of liquid regions, comment on how the overall alloy composition affects the amounts and the compositions of the inclusions.

8.8 Ag–Pb–Zn

The Ag–Pb–Zn system (Figure 8.8) provides a basis for understanding the Parkes process and its various developments which use zinc additions to desilverise lead bullion.

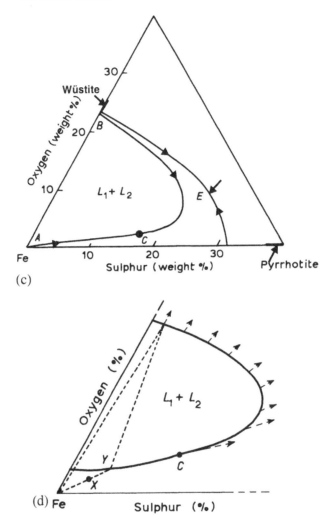

Fig. 8.7 (c) The liquidus projection for the iron-rich portion of the Fe–O–S system.[59] The binary Fe–O system contains a monotectic reaction: L_1 (~0.16 wt.% O) $\rightleftharpoons$ δ–Fe + L_2 (~22.6 wt.% O) (1528°C) and this extends into the ternary system; the liquid miscibility gap is represented by the curve *ACB* where *C* is the 'critical point' occurring at a temperature minimum (~1340°C). Three binary eutectic reactions are shown: $L_2 \rightleftharpoons$ γ–Fe +FeO (wüstite) (1371°C); $L \rightleftharpoons$ γ–Fe + FeS (pyrrhotite) (983°C) and $L \rightleftharpoons$ FeO + FeS. The eutectic valleys meet at a ternary eutectic point $L \rightleftharpoons$ α–Fe + FeO + FeS (920°C). The solubilities of oxygen and sulphur in solid iron are taken to be negligible in the diagram and the equilibria involving the various allostropes of iron are not shown. (d) Schematic view of the liquid miscibility gap (with enlargement of scale in the iron-rich corner) showing the solidification of alloy *X* under non-equilibrium conditions involving the formation of isolated droplets of L_2. The tie-triangle (----) indicates the compositions of Fe, L_1 and L_2 at the beginning of the monotectic reaction. The dashed lines arrowed (−→) represent the changes of composition of individual liquid droplets as solidification proceeds by the separation of iron. The L_1 compositions corresponding to the various L_2 droplets lie along *YC* and are not shown.

Industrial practice dominantly uses both batch processing (normally with several stages of zinc additions), but continuous processing has also been developed.[60] The constituent binary systems are shown in Figures 8.8(a-c). The Ag–Pb system has a eutectic and there is very limited solid solubility. The Ag–Zn system contains primary solid solutions, α (based on Ag) and η (based on Zn) and three intermetallic compounds β, γ and ε which form peritectically. The Pb–Zn system contains a eutectic and there is very limited solid solubility; there is also a monotectic reaction. The liquid miscibility gap associated with the monotectic extends over a substantial part of the ternary system. The boundary of this gap (where the 'dome' intersects the liquidus surface) has a maximum temperature at ~700°C (point C) (Figure 8.8(d)).

The four peritectic reaction curves extending from the Ag–Zn binary intersect the miscibility gap and give rise to invariant monotectic reactions: $L_1 + \alpha \rightleftharpoons L_2 + \beta$ (690°C); $L_1 + \beta \rightleftharpoons L_2 + \gamma$ (660°C); $L_1 + \gamma \rightleftharpoons L_2 + \varepsilon$ (620°C); $L_1 + \varepsilon \rightleftharpoons L_2 + \eta$ (425°C). Tie-lines join the compositions of the respective, L_1 and L_2 points associated with these invariant reactions. The liquidus fields for primary separation of β, γ, ε and η are each divided into two regions by the liquid miscibility gap. These regions lie close to the Ag–Zn binary and Pb-rich corners, respectively.

In the Pb-rich corner of the system. Pb is involved in the following three-phase eutectic reactions (Figure 8.8(e)): $L \rightleftharpoons Pb + \eta$; $L \rightleftharpoons Pb + \varepsilon$ (the reaction curve shows a temperature maximum); $L \rightleftharpoons Pb + \gamma$; $L \rightleftharpoons Pb + \beta$; $L \rightleftharpoons Pb + \alpha$. The Pb-rich corner also includes the three-phase reaction curves that emerge from the miscibility gap involving respectively: L, α, β; L, β, γ; L, γ, ε; L, ε, η. The curve for the composition of liquid in equilibrium with ε and η, is peritectic along its whole length, i.e. $L + \varepsilon \rightleftharpoons \eta$; however, the other curves are reported to be eutectic over part of their length and then to become peritectic in nature.[60] $L + \varepsilon \rightleftharpoons \gamma$; $L + \gamma \rightleftharpoons \beta$; $L + \beta \rightleftharpoons \alpha$. These transitions arise from changes in the liquid compositions relative to the solid phase compositions (see Figure 4.20). The intersections of these curves with the eutectic curves involving Pb give rise to the following invariant reactions. $L + \varepsilon \rightleftharpoons Pb + \eta$; $L + \varepsilon \rightleftharpoons Pb + \gamma$; $L + \gamma \rightleftharpoons Pb + \beta$; $L + \beta \rightleftharpoons Pb + Ag$. (In the above summary of the Pb-rich corner, the subscript 2 has not been applied in designating the liquid phase).

The desilverisation process, which involves intermetallic compound separation from the liquid on cooling, can best be illustrated by reference to the enlarged view of the Pb-rich corner of the system shown in Figure 8.9(e). Zinc is added to molten lead (e.g. an addition of 2% Zn may be made to lead containing 0.2% Ag) and the melt is cooled to just above the melting point of lead. A solid 'crust' is formed, rich in zinc, which contains most of the silver (and gold), together with some lead. Liquid rich in lead precipitates the intermetallic Ag–Zn compound ε as cooling proceeds. The exact solidification sequence depends on the liquid composition but to attain very low silver contents (e.g. < 0.0005%) the final stages of cooling must involve the separation of the zinc-rich solid solution, η Thus, a sequence for alloy X in Figure 8.8(e) would be:

$$L \rightarrow \varepsilon$$
$$L \rightarrow Pb + \varepsilon$$
$$L + \varepsilon \rightarrow Pb + \eta$$
$$L \rightarrow Pb + \eta.$$

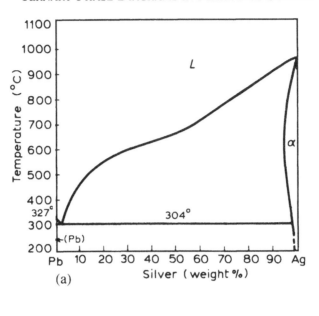

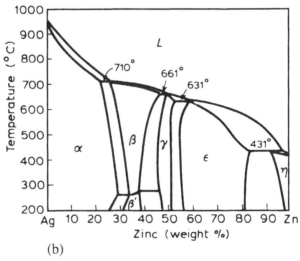

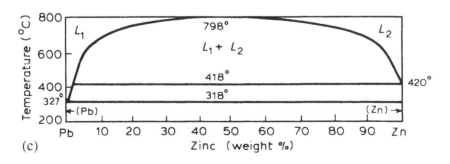

Fig. 8.8 (a) Pb–Ag binary system, (b) Ag–Zn binary system and (c) Pb–Zn binary system.

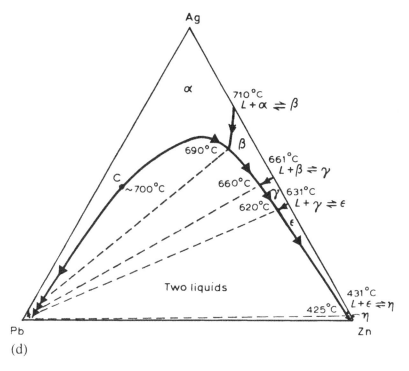

(d)

Fig. 8.8 (d) Liquidus projection of Ag–Pb–Zn system.[60]

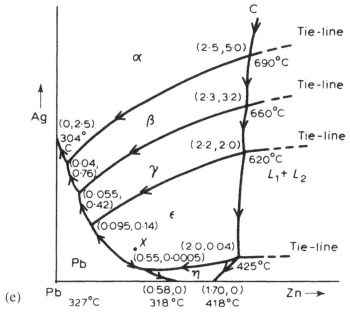

(e)

Fig. 8.8 (e) The details (Semi-schematically) of the Pb-rich corner which are omitted in d: compositions are in wt.% and temperatures are from reference. See also Chapter 10, Ref.32 and Vol.2, O. Kubaschewski.

Whether or not two liquids form during a desilverisation process depends on the amount of zinc added and the temperature. The Parkes process does not involve the formation of two liquids.

However, the Williams continuous process involves two liquids and uses a vessel with a temperature gradient from ~600-650°C at the top to near to the melting point of lead at the bottom. Lead, containing silver, is introduced at the top of the vessel, where a layer of molten zinc is maintained; about 10% Zn can be dissolved by the lead. The precipitation of ε–AgZn occurs as the lead flows downwards through the vessel and is cooled, and desilverised lead is removed. The 'crust' is removed and is treated by distillation to recover the zinc, and by cupellation (selective oxidation) to remove the lead.

Problems
1. Draw, schematically, the part of the isothermal section of the Ag–Pb–Zn system at a temperature just above the melting point of lead (i.e. at ~330°C) showing the equilibria between lead-rich liquid and the Ag-Zn intermetallic compounds γ, ε and η.
2. What is the significance of the invariant reaction $L + \varepsilon \rightarrow Pb + \eta$ in relation to the degree of desilverization than can be achieved in practice?

8.9 Al–Fe–Si, Al–Fe–Mn, Al–Mn–Si and Al–Fe–Mn–Si

Aluminium-based alloys constitute a major group of industrial materials, incorporating some designed for a range of applications, some for as-cast components and others to be used as wrought materials. Age-hardening phenomena provide a key feature of heat treatments in some systems. Among the main alloy elements are copper, magnesium and silicon; the aluminium-rich binary and ternary systems involving these elements are of great interest, e.g. Al–Cu–Mg and Al–Mg–Si. Other elements, such as iron and manganese, are typically found in many commercial alloys and it is necessary, for a full understanding of alloy behaviour, to consider quaternary and other higher order systems. Typically the total alloy content does not exceed about ~10 wt.%, although in the case of Al–Si based alloys, the range of Si contents extends beyond the eutectic composition of ~12 wt.%.

The experimental investigation of phase diagrams of aluminium alloys has been a research theme over most of the 20th century, and by around the mid-century a vast amount of experimental data had been accumulated. In recent years the application of CALPHAD techniques to aluminium based alloys has been a major activity alongside critical assessment of the experimental work reported in the literature. A significant feature of binary and higher order systems is the formation of compounds, based on binary and ternary compositions. e.g. Mg_2Si, Al_2CuMg (S-phase). In many cases the composition range of stoichiometry is very small, and thermodynamic calculations typically assume that the phases are 'line compounds'. The existence of binary and ternary compounds gives rise to systems which include a series of solidification reactions: peritectic and eutectic, including ternary invariant reactions; thus, liquidus projections are often complex.

Three ternary systems are considered here as follows: Al–Fe–Si; Al–Fe–Mn and Al–Mn–Si to provide a basis for illustrating the CALPHAD approach to the modelling of a commercial alloy containing Fe, Mn, Si and Mg as key elements.[*]

8.9.1 Al-Fe-Si

Figures 8.9(a-f)[61, 64] illustrate the main features of the Al-rich part of the system. Ternary invariant solidification reactions occur in the system:

$L + Al_3Fe \rightarrow Al + \alpha AlFeSi$ 629°C
$L + \alpha AlFeSi \rightarrow Al + \beta AlFeSi$ 611°C
$L \rightarrow Al + \beta AlFeSi + Si$ ~577°C
$L + \delta AlFeSi \rightarrow \beta AlFeSi + Si$ 597°C.

A solidification reaction scheme for the Al-rich corner of the systems referring only to the phases and reactions discussed above is shown below. (from Ref. 64. but, temperatures shown below are mainly those quoted by Phillips).[61]

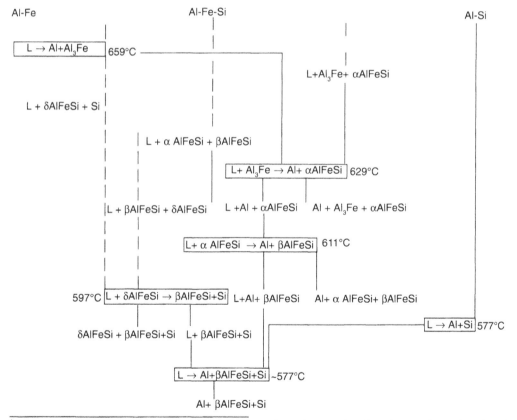

[*] Most of the experimentally determined phase diagrams shown here are based on the experimental work and critical assessments carried out by Phillips.[61] Among other important reviews of earlier work is that by Phragmen.[62] More recently reported experimental and theoretical work and critical reviews (e.g.[63]) have led to various changes in details e.g of temperatures and compositions of invariant reactions. However, the earlier work is the main source used here, giving consistency between binary and ternary data and serving to illustrate the main features.

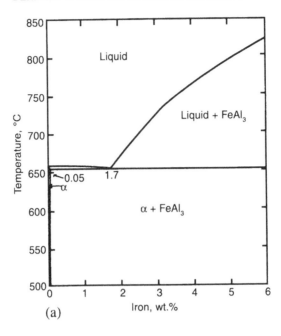

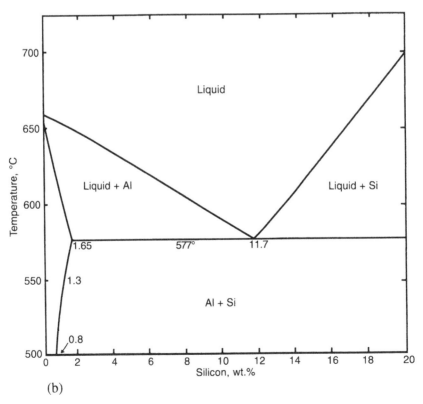

Fig. 8.9 (a) Al-Fe system[61, 66] (Note: the phase Al$_3$Fe[61] is shown as Al$_{13}$Fe$_4$ in Ref. 66) and (b) Al-Si system.[61, 67]

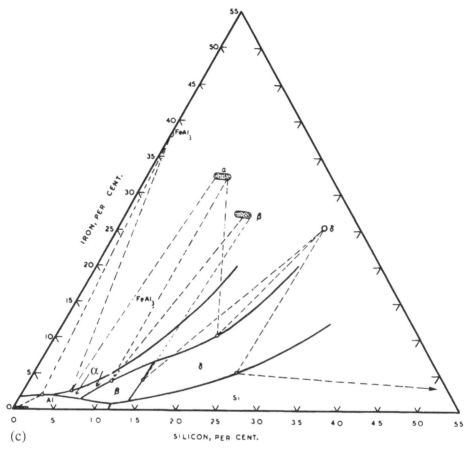

Fig. 8.9 (c) Al–Fe–Si system. General features.[61]

The effect of rapid solidification can be illustrated with reference to the alloy containing 1.8 Fe, 1.1 Si, and Figure 8.9(d).[24] Following the formation of primary Al dendrites, the liquid composition moves on to the Al + Al$_3$Fe eutectic valley. However, whereas in slow cooling the eutectic mixture should begin to solidify, in the case of rapid solidification undercooling occurs and the liquid composition continues to move on the extended Al liquidus surface to meet the metastable extension of the $L \rightarrow$ Al + αAlFeSi eutectic curve. In the case of a *DC* ingot, the cooling rate is such that solidification proceeds to completion by this metastable eutectic. With the faster cooling rate associated with a twin-roll Hunter process, this eutectic does not form, but the undercooling proceeds further, along the metastable Al liquidus to reach the metastable extension of the eutectic: $L \rightarrow$ Al + βAlFeSi Thus, in neither of the rapidly solidified cases does the equilibrium Al$_3$Fe compound form. An estimate of the non-equilibrium solidification path can be

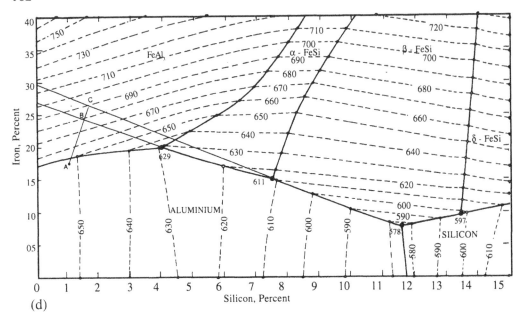

(d)

Fig. 8.9 (d) Liquidus projection.[61] Note: The letters *ABC* refer to the solidification path of an alloy of composition *A* (1.8 Fe, 1.1 Si wt.%) under rapidly solidified conditions. *B* is the liquid composition on the metastable extension of the *L* → Al + αAlFeSi (designated α-FeSi in the figure) eutectic curve, and C is the liquid composition on the metastable extension of the Al + βAlFeSi (designated β-FeSi in the figure) eutectic curve. Note: The temperature of the ternary eutectic L → Al + β-FeSi + Si is shown as 578°C by Phillips,[61] while the binary eutectic L → Al + Si is shown as 577°C (Fig. 8.9b). However Ref.64 shows the ternary eutectic temperature as below 577°C.

obtained using the Scheil equation independently for Fe and Si (equilibrium *k* values 0.02 and 0.13 respectively); the volume fractions of primary Al dendrites for the two cooling rates considered above are calculated[24] as approximately 0.3 and 0.35 respectively.

Problem
    Al–Fe–Si system. (Figures 8.9(d and f)). Consider an alloy with 1.8 Fe and 1.1 Si (wt.%). The solidus projection shows the position of the alloy as located in the Al + αAlFeSi region in the solid state under equilibrium conditions. By reference to the liquidus and solidus projections (Figures 8.9(d and f)) describe the equilibrium solidification sequence of this alloy.

8.9.2 Al–Fe–Mn

Figures 8.9(g-i) illustrate the main features of the Al-rich part of the system.[61]

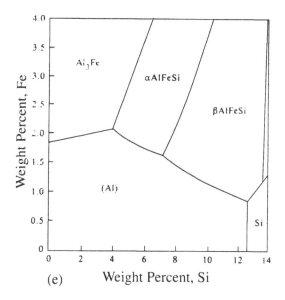

(e)   Weight Percent, Si

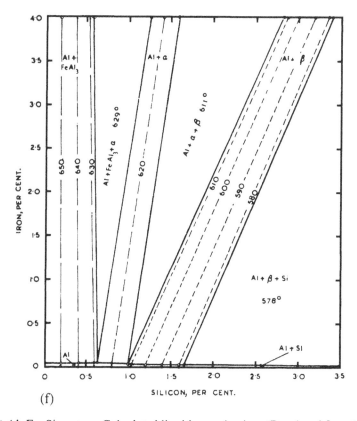

(f)

Fig. 8.9 (e) Al–Fe–Si system. Calculated liquidus projection. (Reprinted from N. Saunders and A.P. Miodownik. *CALPHAD, Calculation of Phase Diagrams*, 1996, 324, with permission from Elsevier Science) and (f) Al–Fe–Si system. Solidus projection.

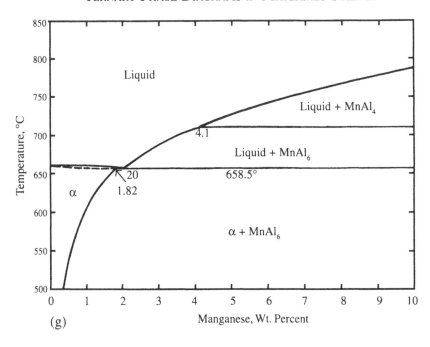

(g)

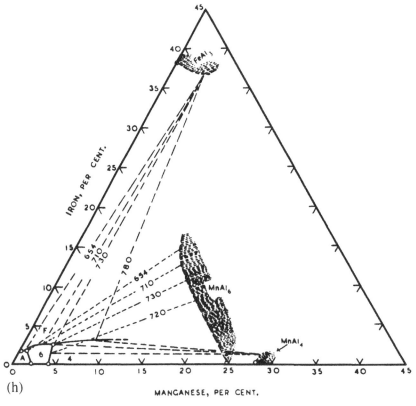

(h)

Fig. 8.9 (g) Al–Mn system[61, 68] and (h) Al–Fe–Mn system. General features.

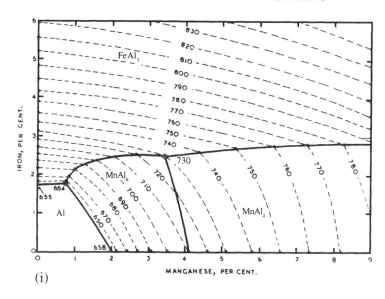

(i)

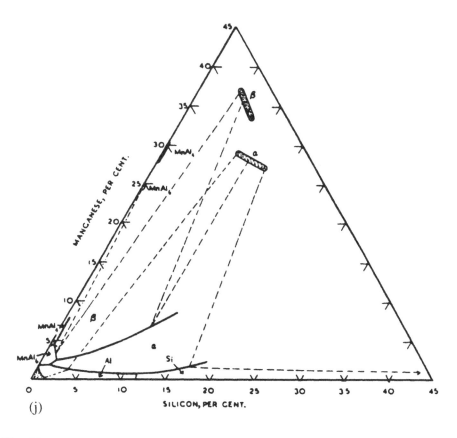

(j)

Fig. 8.9 (i) Al–Fe–Mn system. Liquidus projection[61] and (j) Al–Mn–Si system. General features.

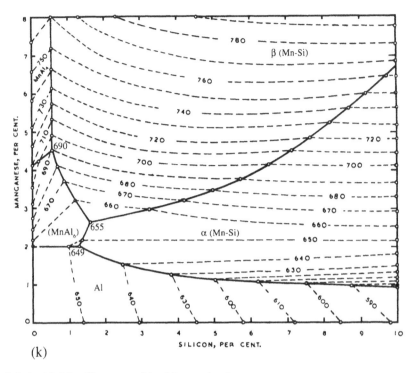

(k)

Fig. 8.9 (k) Al–Mn–Si system. Liquidus projection.

Problem

Al–Fe–Mn system: (Figures 8.9(g, h, i)). From the information provided for the relevant binary systems, and that for the ternary system state the nature of the ternary invariant solidification reactions.

8.9.3 Al–Mn–Si

Figures 8.9(j and k) illustrate the main features of the Al-rich part of the system.[61,65]

Problem

Al–Mn–Si system: (Figures 8.9(j and k)). State the nature of the ternary invariant solidification reactions shown in Figures 8.9(j and k).

8.9.4 Al-Fe-Mn-Si

Consideration of the Al–Fe–Mn–Si quaternary system[13] provides a good example of the way in which CALPHAD can be used to model commercial aluminium alloys; in this case the alloy providing the example is AA3004, used for thin walled containers, whose composition can be typified as a five component material, Al–1Mn–1.2Mg-0.5Fe-0.2Si (wt.%). As a background to discussing this example, it is appropriate to note a few general points concerning the CALPHAD approach to aluminium alloys. Although the solidification and solid state reactions are complex in the ternary and higher order aluminium alloys, modelling of aluminium alloys benefits from the fact that few intermediate phases exhibit wide ranges of stoichiometry. Of the many binary and ternary intermediate phases, a substantial number of these tend to be stoichiometric in nature e.g. Mg_2Si and Al_2CuMg. In cases where there is substantial solubility, e.g. for Al_6Mn and $\alpha AlFeMnSi$, the transition metals basically mix on one sub-lattice while Si mixes on the Al sub-lattice and the phases can then be dealt with as conventional line compounds. A further helpful feature is that most of the alloying elements encountered in commercial Al alloys show only limited solubility in the Al-based solid solution that forms the matrix of the alloys.

For the AA3004 alloy, the relevant regions of the binary phase diagrams: Al–Si, Al–Mn and Al–Fe are shown in Figures 8.9(a, b and g). The binary compounds to be considered are: Al_6Mn, Al_4Mn and Al_3Fe. In the Al–Fe–Mn system (Figures 8.9(h and i)) there is a substantial solubility of Fe in the Al_6Mn phase and a small solubility of Mn in Al_3Fe. The addition of Si to Al–Mn and Al–Fe leads to the formation of various ternary intermediate phases, but although there is significant mixing between Al and Si in the $\alpha AlMnSi$ compound, the other phases can be treated accurately as line compounds. When the ternary systems Al–Fe–Mn and Al–Fe–Si are joined to form the quaternary Al–Fe–Mn–Si system, the $\alpha AlMnSi$ phase is found to extend almost completely to the Al–Fe–Si ternary. To incorporate the effect of Mg, account is taken of the Mg_2Si phase from the Al-Mg-Si system.

Figures 8.9(l and m)[13] shows the calculated plot of phase% vs temperature for the alloy. Solidification begins with the formation of primary Al, followed soon by the formation of Al_6Mn; there follows a peritectic reaction involving $\alpha AlFeMnSi$, which partly consumes the Al_6Mn. The proportion of $\alpha AlFeMnSi$ increases as the alloy cools below the solidus, and it becomes the dominant intermetallic phase in the solid state just below 600°C. However, the alpha phase disappears at around 400°C, as Si is taken up by the formation of Mg_2Si, which acts as the precipitation hardening phase.

8.10 Ti-Al BASED ALLOYS

The past half century has seen the development of a wide-ranging group of industrially important titanium based alloys, providing a variety of combinations of properties including relatively low density and good strength. The aerospace industry has been one of the most significant beneficiaries from the availability of titanium alloys, while major applications have also been found in many other fields. Titanium shows an allotropic transformation from the high temperature, bcc β form to the low temperature cph α form

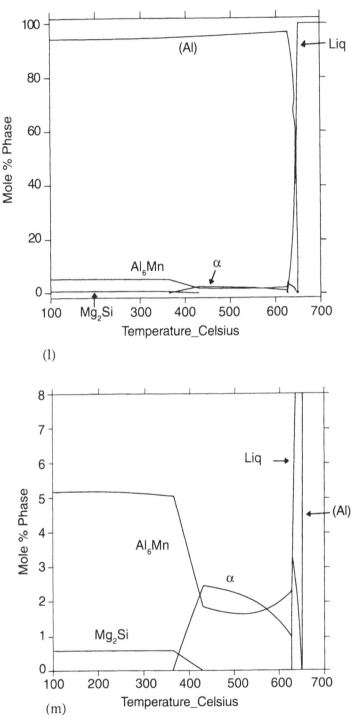

Fig. 8.9 AA 3004 alloy. (l) Calculated mole % phase vs temperature plot and (m) Expanded version of Figure 8.9(l). (Reprinted from N. Saunders and A.P. Miodownik, *CALPHAD, Calculation of Phase Diagrams*, 1996, p.326 with permission from Elsevier Science).

at 882.5°C and the behaviour and design of titanium alloys is based on the effect of alloying additions on this transformation. Of the wide range of elements relevant to the alloying of titanium, oxygen is always present, entering into interstitial solid solution in both the α and β phases. Nitrogen and carbon, which are present as impurity elements, also enter interstitial solution. All CP (commercial purity) grades of titanium depend on alloying by oxygen to provide strength. Also many commercial alloys have some oxygen added to increase strength. However, to avoid reduction of ductility, the oxygen content must be carefully controlled during processing; typically the aim is to limit the oxygen content to < ~1200 ppm.

Alloying elements are commonly classified into three main types in relation to their effects on the β–α transition.

i. Certain elements are referred to as α stabilisers; these raise the allotropic transformation temperature, and thus extend the range of temperature and composition over which the α phase exists. Oxygen is an important example of this type; another example, aluminium, is of critical importance in many types of titanium alloy. The Ti-Al phase diagram, as shown in Figure 8.10(a),[69] illustrates the α-stabilising effect and also includes other features such as the formation of intermetallic compounds, which are discussed below.

ii. β stabilisers act to lower the allotropic transformation and provide scope for achieving extensive ranges of temperature and composition over which the β phase exists. In some systems such as Ti–V (Figures 8.10(b))[70, 71] and Ti–Nb, the addition of a sufficiently large proportion of the alloy elements extends the range of β stability to low temperatures. There are other important cases where the β phase undergoes a solid state transformation such as a eutectoid: $\beta \rightarrow \alpha$ + an intermetallic compound; e.g. in the Ti–Si system where the compound is Ti_5Si_3.

iii. So-called 'neutral' elements have only a relatively small effect on the allotropic transformation temperature. Important examples are Sn and Zr; the latter element shows the same type of β–α transformation as does Ti, and the Ti–Zr system shows a complete range of β and of α solid solutions.

Industrially used titanium based alloys, of which there is a great variety, range from binary (CP Ti, containing O, but neglecting the presence of N and C) through ternary compositions (+ oxygen), to multicomponent systems. The most widely used alloy, since its development some 50 years ago is the ternary composition, Ti–6Al–4V (wt.%) with structures involving both α and β solid solutions. A multicomponent example is the alloy designated IMI 834: Ti–5.8 Al–4 Sn–3.5 Zr–0.7 Nb–0.5 Mo–0.35 Si–0.06 C (wt.%); this alloy, which consists mainly of α phase, is representative of a group which has been developed to possess good creep and oxidation resistance at temperatures up to ~ 630°C, for use in components such as compressors for aero jet engines.

Phase diagram information, ranging from binary to multicomponent systems, is clearly of vital importance in the design and application of titanium alloys. A huge amount of experimental work has been carried out on phase equilibria and transformations. One of the difficulties is the effect of oxygen which can move phase boundaries significantly at small concentrations. This may, in part, account for the fact that around 50 versions of

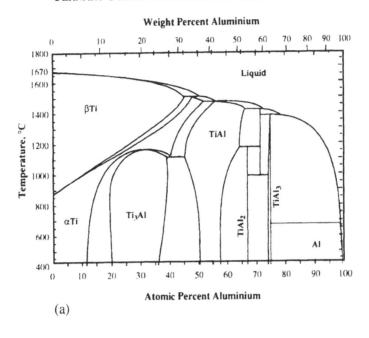

(a)

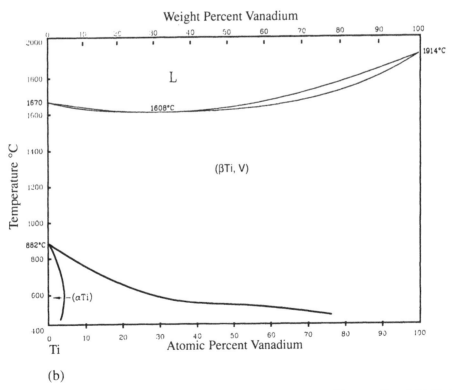

(b)

Fig. 8.10 (a) A recent version of the Ti–Al phase diagram, due to Okamoto[69] and (b) Ti–V phase diagram.[70, 71] (Reproduced with permission from ASM International, Materials Park, OH 44073-0002, USA).

the Ti–Al diagram have been presented, albeit many of the recent versions differ only in minor detail. The CALPHAD approach has also been successfully applied, e.g.,[72] with oxygen being considered as an important case.

Considering here the significance of ternary phase diagrams, reference can be made to systems containing both an α stabiliser and a β stabiliser, e.g. Al and V. It is helpful to consider a schematic vertical section (isopleth) through such a system, (Figure 8.10(c))[73] bearing in mind that the α–β tie-lines do not lie within the plane of the section. Figure 8.10(c) shows the α and β transus curves as full lines, together with a dashed curve which represents the metastable situation when β is cooled rapidly, leading to a martensitic type of transformation of β → α'; the start and finish transformation temperatures for martensite (M_s and M_f) are typically separated by only a narrow temperature range and are shown as a single line in the Figure.

Figure 8.10(c) also indicates a possible classification for titanium alloys on the basis of the proportions of the constituent phases; α and β. Brief comments on some of the main features of the various classes of alloys are presented below, bearing in mind that thermomechanical treatments under a variety of conditions are available to control the types, proportions and morphologies of the phases. It should be noted that in schematic terms the diagram in Figure 8.10(c) can be used generically to represent multicomponent systems containing a number of both α and β stabilisers in various proportions.

i.   The α type of alloy is limited to compositions in which the retention of β (even in metastable form) is not possible, and such alloys are effectively 'non heat-treatable'.

ii.   Near-α alloys contain only small proportions of β stabilisers and are processed either in the β phase field of the system or in the α + β field e.g. as with IMI 834 alloy to give a predominantly α phase microstructure for optimum creep resistance.

iii. The α–β type of alloy, of which the 'classic' case is Ti–6Al–4V, covers a wide range of compositions and temperatures involving both types of phase; the relative proportions of the phases are strongly influenced by the thermomechanical processing conditions. If the alloy is processed by heating to a temperature relatively high in the α + β field, at which mechanical working may be applied, the resultant β phase composition and the cooling rate following the thermomechanical treatment will obviously influence the final structure of the alloy at room temperature. For example, if the β composition lies to the right of the point at which the M_s/M_f curve reaches room temperature then rapid cooling will lead to the retention of metastable β. For solute contents to the left of the M_s/M_f curve intersection with room temperature the β can transform to martensite, α'. If cooling following thermomechanical treatment is slow the β phase can transform to varying extents to α phase. The formation of the metastable omega, ω, phase can also occur when retained β is aged at relatively low temperatures (typically below 450°C).

iv.   In near-β alloys the concentration of β stabilising solutes is high enough to lower the M_s/M_f curve to below room temperature. Such alloys can be heat treated in the α + β field, when a variety of precipitation processes can occur, including the formation of metastable ω phase.

v.   Concerning β type alloys, their very high solute contents lead to the β phase being weakly metastable at room temperature so that decomposition into α + β does not occur during industrial thermomechanical treatments.

In recent years, intense interest (e.g.[74]) has developed in the intermetallic compounds Ti_3Al (designated $α_2$) and TiAl (designated γ) in the Ti–Al system (Figure 8.10(a)). Interest focuses on the potential of using these compounds as the basis for developing alloys capable of use at temperatures in excess of those achieved by the near-α type. Thus, creep and oxidation resistance, initially up to ~700°C, are being sought. At room temperature both of the compounds are very limited in their ductility, and one of the objectives of alloying is to ameliorate this problem. Currently the emphasis is on the TiAl based alloys, as offering the best potential, and alloy development and design have particularly focused on the addition of transition elements such as Nb, V, Mo, Cr, Mn, and Ta, often in small proportions.

The presence of substantial proportions of such elements leads to the formation of additional phases in the alloys with potentially beneficial effects. Notably, for example, an ordered cubic B2 β phase (sometimes designated $β_o$ or $β_2$) forms in certain composition ranges in stable form, and over wider ranges in metastable form, after quenching from high temperatures. Also of considerable potential interest is an orthorhombic ternary phase, designated O, based on Ti_2AlNb; this phase can be equilibrated with the $α_2$, γ and $β_o$ phases. Figures 8 10(d and e) and illustrate some of these features, in the Ti-Al-V and Ti-Al-Nb systems.[74-76] O phase does not form in Ti-Al-V alloys and details of the equilibria in the Ti-Al-Nb system have yet to be fully and definitively established, including the influence of oxygen.

There is currently considerable interest in the directional solidification of TiAl (γ) based alloys (e.g. Ref. 77). This relates particularly to the formation of the lamellar $α_2$ + γ microstructures produced during cooling in the solid state. The mechanical properties of the alloys are strongly dependent on the orientation of the lamellae with respect to the loading axis. In order to achieve optimum property combinations processing by directional solidification techniques has been used to produce microstructures with the lamellae aligned parallel to the direction of crystal growth during solidification. However, for reasons relating to the phase equilibria during solidification and subsequent solid state cooling, there are difficulties in obtaining the required microstructures using binary TiAl alloy compositions. Accordingly attention has been turned to TiAl compositions with small additions of, for example, of Si or Mo. Among the approaches adopted (e.g. Ref. 77) is the use of a seeding crystal and a zone melting technique, with the seed crystal composition being selected in the light of solidification sequences as based on experimental work and interpreted from the relevant ternary liquidus projection.

Problems
1.   Consider the isothermal section at 700°C of a hypothetical ternary system Ti–Al–X, where X is a beta stabilising element in the range (wt.%) up to ~20% Al and 25% X. Assume the following composition data for phase fields in the system.

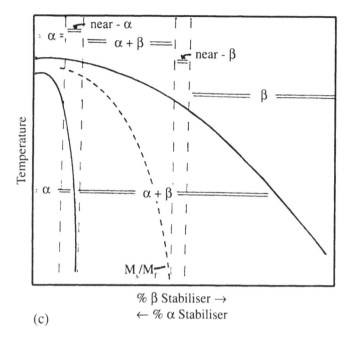

(c)

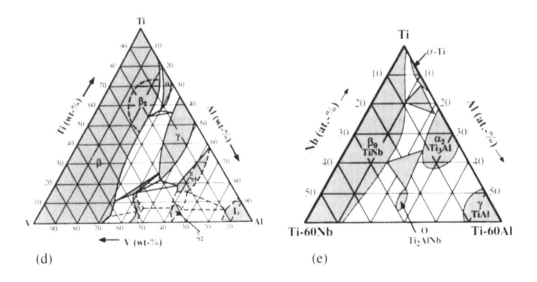

(d)

(e)

Fig. 8.10 (c) Schematic quasi-vertical section for titanium alloys containing both α and β stabilising elements,[73] (d) Ti-Al-V system; isothermal section at 1473 K, showing extent of ordered B2 (designated β_2) field. Ref. 73, adapted from Ref. 75 and (e) Ti-Al-Nb system; isothermal section at 1173 K, indicating the presence of ternary O-phase, based on Ti_2AlNb and three-phase equilibria between O, α_2 and B2 (designated β_o); Ref. 74, adapted from Ref. 76.

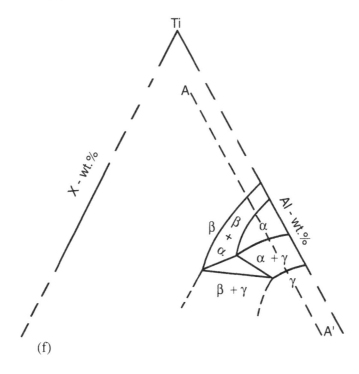

(f)

Fig. 8.10 (f) Ti–Al–X system. Schematic part of isothermal section at a temperature just below that of a Class II peritectic invariant reaction involving L, α, β and γ.

A three-phase field, $\alpha + \alpha_2 + \beta$ exists, with the compositions (wt.%) of the three phases as follows:

	%Al	%X
α	9.5	3
α_2	14.5	3.5
β	7	12

Binary systems phase boundary compositions (wt.%):

Ti–Al	α 9%Al	α_2 14% Al
Ti–X	α 1.5% X	β 8% X

Use the above data to draw a possible form of the isothermal section in the composition range stated, assuming that the only phases present are α, α_2 and β.

2. Consider an alloy system Ti–Al–X, where X is a β stabilising element. Part of the Ti-rich region of the system is shown in Figure 8.10(f) in the form of an isothermal

section at a temperature just below that of an invariant reaction of the Class II peritectic type, involving the phases: liquid, β, α and γ.

Draw schematically a possible form of a liquidus projection for the part of the system shown in the Figure and show the configuration of the tie-triangles at the invariant temperature. Also draw schematically a possible form of a vertical section at the constant content of β stabiliser shown by the dashed line in the Figure.

8.11 Al_2O_3–MgO–SiO_2 (corundum–periclase–cristobalite/tridymite)

This ternary system (Figures 8.11(a-d)) is important in relation to various ceramics[78, 79] non-metallic inclusions in steels,[80] slags[49] and also in petrology.[10, 81]

It contains binary and ternary compounds as follows:

$MgO \cdot Al_2O_3$ (spinel)   :   Congruently melting   ~2135°C

$3Al_2O_3 \cdot 2SiO_2$ (mullite)   :   Congruently melting   ~1850°C

$2MgO \cdot SiO_2$ (forsterite)   :   Congruently melting   ~1900°C

$MgO \cdot SiO_2$ (enstatite)   :   Incongruently melting   1557°C

(There are three modifications of MgO–SiO_2: protoenstatite (the stable high temperature form >1040°C); enstatite (the stable low temperature form); clinoenstatite (can exist as a metastable form at room temperature)).

$4MgO \cdot 5Al_2O_3 \cdot 2SiO_2$ (sapphirine) :   Incongruently melting   1482°C

$2MgO \cdot 2Al_2O_3 \cdot 5SiO_2$ (cordierite)   :   Incongruently melting   1465°C

Of these compounds spinel, mullite and cordierite show ranges of solid solution at certain temperatures, that of spinel being particularly extensive, (from ~60–80% Al_2O_3). These maximum ranges are shown by the hatched lines in Figure 8.11(d) and there are corresponding extensive two-phase regions in the solid state. Other features to note in the system include the liquid miscibility gap based on the MgO–SiO_2 system and the existence of cristobalite and tridymite as allotropic forms of SiO_2.

The Al_2O_3–MgO–SiO_2 system is an example of relatively stable oxides containing cations, which appear in only one predominant state of oxidation. In this system the gas phase can usually be ignored. An example of a ternary system, FeO–O–Si, where the gas phase is important is discussed in Section 8.6. For an example of a ternary system where the gas phase is important reference may be made to the Fe–O–Si system.[10]

Invariant reactions are as follows:

Binary Systems (Figures 8.11(a-c))

MgO–Al_2O_3	~1850°C	$L \rightleftharpoons$ MgO (periclase) + spinel
	~1925°C	$L \rightleftharpoons Al_2O_3$ (corundum) + spinel
Al_2O_3–SiO_2	~1840°C	$L \rightleftharpoons Al_2O_3$ + mullite
	~1590°C	$L \rightleftharpoons SiO_2$ (cristobalite) + mullite
MgO–SiO_2	~1860°C	$L \rightleftharpoons$ MgO (periclase) + forsterite
	1557°C	$L +$ forsterite $\rightleftharpoons$ protoenstatite

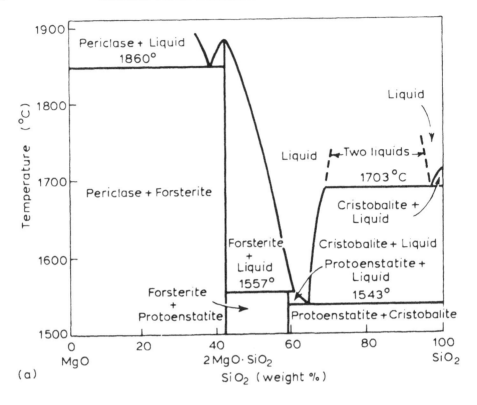

(a)

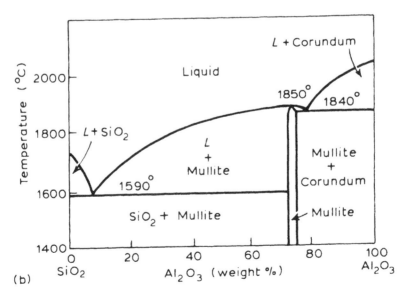

(b)

Fig. 8.11 (a) MgO–SiO$_2$ binary system and (b) SiO$_2$–Al$_2$O$_3$ binary system. (A revised version[83] of the Al$_2$O$_3$–SiO$_2$ system shows mullite as melting incongruently (peritectic reaction: L + Al$_2$O$_3 \rightleftharpoons$ mullite). However for purposes of illustration the congruently melting case has been retained here).

1543°C $\quad L \rightleftharpoons$ protoenstatite + SiO_2 (cristobalite)

1703°C $\quad L_1 \rightleftharpoons L_2 + SiO_2$ (cristobalite)

Ternary Compositions

1578°C $\quad L +$ Al_2O_3 (corundum) $\rightleftharpoons$ spinel + mullite

1482°C $\quad L +$ spinel + mullite $\rightleftharpoons$ sapphirine

1460°C $\quad L +$ mullite $\rightleftharpoons$ sapphirine + cordierite

1453°C $\quad L +$ sapphirine $\rightleftharpoons$ spinel + cordierite

1440°C $\quad L +$ mullite $\rightleftharpoons$ cordierite + SiO_2 (tridymite)

1355°C $\quad L \rightleftharpoons$ SiO_2 (tridymite) + cordierite + protoenstatite

1370°C $\quad L +$ spinel $\rightleftharpoons$ forsterite + cordierite

1365°C $\quad L \rightleftharpoons$ forsterite + protoenstatite + cordierite

1710°C $\quad L \rightleftharpoons$ MgO (periclase) + forsterite + spinel

Note also the 'saddle' points

1720°C $\quad L \rightleftharpoons$ forsterite + spinel

1367°C $\quad L \rightleftharpoons$ protoenstatite + cordierite

1465°C $\quad L +$ mullite $\rightleftharpoons$ cordierite

Note on Terminology

Having considered here a detailed example of a ceramic system, it is appropriate to mention certain features of the terminology which are sometimes applied to phase diagrams in ceramics and in geology[10, 49] and which differ from the usual metallurgical terminology. Curves representing the compositions of liquid in equilibrium with two other phases and which subdivide the liquidus into primary fields may be called *boundary curves*. With reference to solidification they may be described as either *subtraction curves*, along which crystalline phases separate from the liquid (corresponding to eutectic solidification, e.g. $L \to \alpha + \beta$) or *reaction curves*, along which crystalline phases both precipitate and dissolve (peritectic solidification, e.g. $L + \alpha \to \beta$).[10]

The term *Alkemade line* (or compatibility line) refers to a straight line joining the composition points of two co-existing phases whose primary phase fields share a common boundary curve. For example, in Figure 6.2 the primary fields of *A* and *C* share the boundary curve *ji*, so that *AC*, the side of the composition triangle is an Alkemade line; the primary fields of *X* and *C* share the boundary curve *ihg* so that *XC* (the quasi-binary section) is an Alkemade line. This system (Figure 6.2) contains five boundary curves and there are correspondingly five Alkemade lines *AX*, *XB*, *BC*, *AC*, *XC*. A liquid whose original composition lies on an Alkemade line will solidify to a mixture of the two solid phases whose compositions lie at the ends of the line.

Alkemade's theorem states that the intersection of an Alkemade line with its pertinent boundary curve represents a temperature maximum for the boundary curve, e.g. in Figure 6.2 points *d, e, f, j, h*. If the boundary curve (or any tangent drawn from it) intersects an extension of the pertinent Alkemade line, but not the actual line then that part of the boundary curve is of the reaction type. The point of intersection again represents a temperature maximum for the boundary curve, e.g. in Figure 6.11 a tangent drawn to reaction curve *dh* intersects the extension of line *AX*.

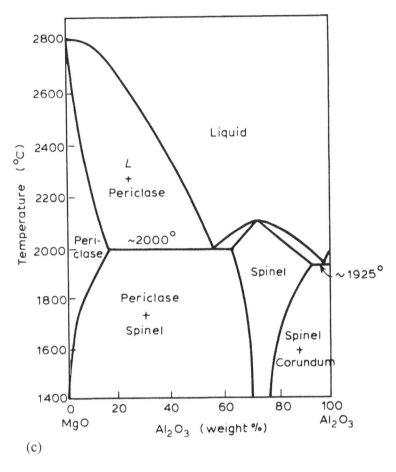

(c)

Fig. 8.11 (c) MgO–Al$_2$O$_3$ binary system. (Note: The value of ~2000°C shown for the temperature of the $L \rightleftharpoons$ periclase + spinel reaction is taken from a more recent source[79] than that used for the value of ~1850°C shown in (d)).[56]

Concerning ternary invariant reactions, the types designated peritectic (II) and (III) on p.63 can be described in terms of the points where the reaction curves intersect as *tributary reaction points and distributary reaction points* respectively.[10] For each invariant reaction there is an *Alkemade triangle* (compatibility triangle), formed from three Alkemade lines.

The following procedure may be followed in approaching the interpretation of a complex ceramic system. Identify the Alkemade lines (one for each boundary curve). Find the nature of the boundary curves by determining whether the tangents intersect the pertinent Alkemade line, or its extension (cf. Figure 4.17). Taking into account the slopes of the boundary curves identify the nature of the reaction points (p.104). The application of this approach to complex systems such as Al_2O_3–MgO–SiO_2 and Al_2O_3–CaO–SiO_2 is described by Ehlers.[10]

Problems
1. Draw an isothermal section representing the solid state constitution for the Al_2O_3–MgO–SiO_2 system.
2. By reference to Figure 8.11(d), describe the solidification sequences of the following ceramic compositions (wt.%), assuming equilibrium conditions.
    (a) $15MgO$–$5SiO_2$ consisting of spinel + mullite when solidification is completed.
    (b) $22MgO$–$7SiO_2$ consisting of spinel + sapphirine when solidification is completed.
    (c) Sapphirine composition consisting of sapphirine only when solidification is completed.
    (d) $8MgO$–$17Al_2O_3$ consisting of SiO_2 + cordierite when solidification is completed.
3. Compare the melting characteristics of the following ceramic compositions (i.e. the variation of liquid content with temperature within the melting range) and comment on the relevance to the firing behaviour.
    (a) 45% SiO_2–4% Al_2O_3 (forsterite type)
    (b) 65% SiO_2–5% Al_2O_3 (steatite type)
    (c) 58% SiO_2–24%Al_2O_3 (cordierite type)
4. In the context of the fluxing effect of MgO on Al_2O_3–SiO_2 fire-clay bricks, describe the result of adding 1% MgO to a material of composition 42% Al_2O_3, 58% SiO_2, heated to equilibrium at 1460°C (refer to Figure 8.11(e)).

8.12 BaO–CuO–YO$_{1.5}$

The compound $YBa_2Cu_3O_{7-x}$(represented as the 123 compound) in the BaO-CaO-YO$_{1.5}$ (BaO-CuO-$1/2Y_2O_3$) system is a high temperature superconductor with a critical temperature, T_c, of 92 K. Many investigations have been made of the phase equilibria, which, among other factors, depend on the oxygen pressure of the system, but the literature shows significant differences between the results of different researchers e.g. in relation to the temperatures and compositions of the invariant reactions. The information shown here is intended to be representative of the important features of the system in relation to the processing, both of single crystals and of bulk polycrystalline samples; sintering, involving liquid phase reactions, is the route adopted and choice of temperature and of material composition are critical factors for achieving optimal results for the 123 compound. Thus phase diagram information is essential (e.g. Ref. 84-86) and in this context Figures 8.12(a-c) show important features of the phase equilibria, typical of conditions of 0.2 atm. of oxygen i.e. in air.

A complication to be noted, but not taken account of here, is the effect of the presence of water and carbon dioxide. For example, barium-rich ternary oxides have a high affinity for water and carbon dioxide, so that when phase diagram investigations are carried out

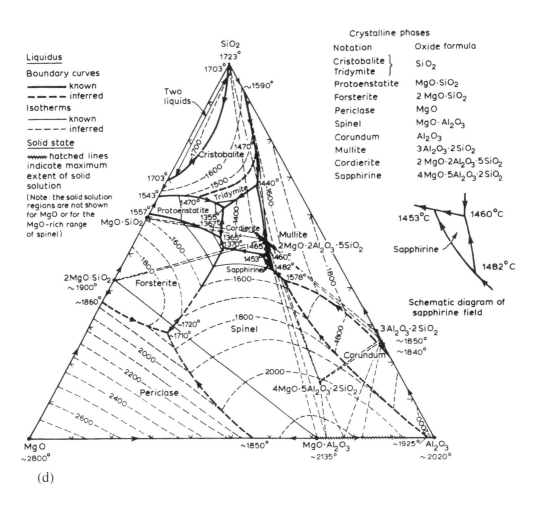

(d)

Fig. 8.11 (d) Composite representation of Al_2O_3–MgO–SiO_2 system, showing liquidus projection and boundaries of phase regions in the solid state. The latter includes information (hatched lines) on ranges of solubility for spinel (MgO · Al_2O_3), mullite ($3Al_2O_3$ · $2SiO_2$) and cordierite (2MgO · Al_2O_3 · $5SiO_2$); however, solid solution regions are not shown for MgO or for the MgO or for the MgO-rich range of spinel. The limits of the solid solubility ranges are markedly temperature dependent.[56] Note: A three-dimensional representation (anaglyph) of the Al_2O_3–MgO–SiO_2 system is given in Ref. 82. (Reproduced by permission of The American Ceramic Society, P.O. Box 6136, Westerville, Ohio 43086-6136. Copyright 1960 and 1968 by The American Ceramic Society. All rights reserved).

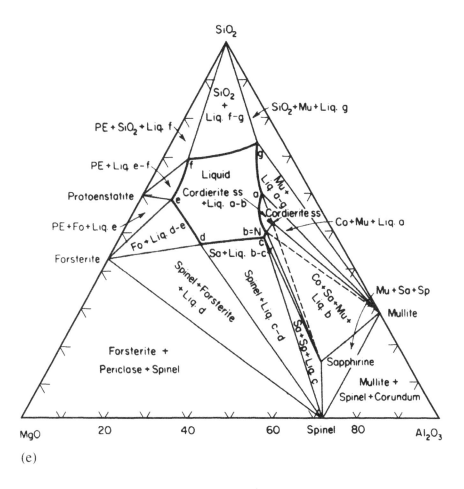

(e)

Fig. 8.11 (e) Isothermal section of Al₂O₃–MgO–SiO₂ system at 1460°C.[78, 81] The range of solid solutions for cordierite is shown as a heavy line but solid solution ranges for other compounds are not represented. The invariant reaction (at point N): liquid (b) + mullite ⇌ sapphirine + cordierite occurs at 1460°C and is represented by the four-phase region liquid (b) + mullite + sapphirine + cordierite; the dashed lines within this region connect the pairs of phases involved. Abbreviations used: cordierite (Co) forsterite (Fo), mullite (Mu), protoenstatite (PE), sapphirine (Sa), spinel (Sp), liquid (Liq.). (Reproduced from W. Schreyer and W. Schairer, *Journal of Petrology*, 1961, **2**(3), 361 with permisssion from Oxford University Press).

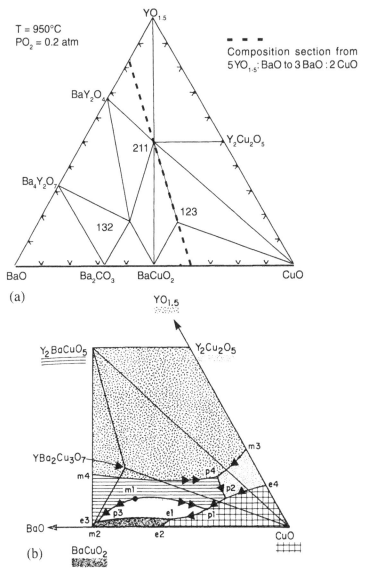

(a)

(b)

Fig.8.12 BaO–CaO–YO$_{1.5}$ system. 0.2 atm. partial pressure of oxygen. (a) Isothermal section (mole.%) representing the solid state compatibility triangles at a temperature just below the lowest melting eutectic in the system i.e below ~890°C. Of the binary compounds from the three constituent binary systems, the following are relevant for the present purposes: BaCuO$_2$; Y$_2$Cu$_2$O$_5$; also two of the ternary compounds are particularly relevant: Y$_2$BaCuO$_5$ (211) and YBa$_2$Cu$_3$O$_{7-x}$ (123). None of the components of the system, or these particular compounds are reported as possessing significant solid solubility. The compounds Ba$_2$Y$_2$O$_5$ and Ba$_3$Y$_4$O$_9$ (shown by some workers) and their associated equilibria,are omitted from the diagram[86] adapted from (Ref. 84) and (b) Liquidus projection showing primary phase fields in the YO$_{1.5}$–BaCuO$_2$–CuO region; determined in air at 0.84 atm. total pressure; for approximate temperatures of invariant reactions see Table 1. (Adapted from Ref. 87 by Ref. 88. (With permission from Wiley-VCH, STM Copyright and licensees).

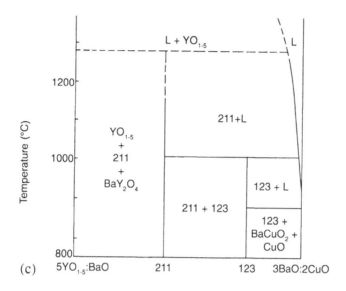

Fig. 8.12 (c) Simplified version of vertical section through the system along the compositional line from $5YO_{1.5}$: BaO to $3BaO : 2CuO$, which incorporates the line joining the 211 and 123 compounds. (adapted from M. Murakami et al.[89]) (Reprinted with permission of the American Ceramic Society, P.O. Box 6136, Westerville, OH 43086-6136, © 1998 American Ceramic Society. All rights reserved). Note: Ref. 84, which represents the results of various workers includes a more complex form of this vertical section for the regions above 1200°C involving liquid, yttria and 211. Also in the area labelled above as 123 + L these are in fact, several regions derived from the relationships associated with the ternary eutectic: $L \rightarrow BaCuO_3$, 123 and CuO. Binary invariant and ternary invariant reactions involving liquid and solid phases relevant to the present discussion are shown below.

Reaction	Invariant Point	Approx. Temp°C
$L \rightleftharpoons YBa_2Cu_3O_{7-x}$ (123) $+ BaCuO_2 + CuO$	e_1	890
$L \rightleftharpoons BaCuO_2 + CuO$	e_2	920
$L + Y_2BaCuO_5$ (211) $\rightleftharpoons YBa_2Cu_3O_{7-x}$ (123) $+ CuO$	p_1	940
$L + Y_2Cu_2O_5 \rightleftharpoons Y_2BaCuO_5$ (211) $+ CuO$	p_2	975
$L \rightleftharpoons Y_2BaCuO_5$ (211) $+ BaCuO_2$	e_3	~1000
$L + Y_2BaCuO_5$ (211) $\rightleftharpoons YBa_2Cu_3O_{7-x}$ (123) $+ BaCuO_2$	p_3	~1000
$L + Y_2BaCuO_5$ (211) $\rightleftharpoons YBa_2Cu_3O_{7-x}$ (123)	m_1	1015
$L \rightleftharpoons BaCuO_2$	m_2	1015
$L \rightleftharpoons Y_2BaCuO_5$ (211) $+ CuO$	e_4	1110
$L + YO_{1.5} \rightleftharpoons Y_2Cu_2O_5$	m_3	1122
$L + YO_{1.5} \rightleftharpoons Y_2BaCuO_5$ (211)	m_4	1270
$L + YO_{1.5} \rightleftharpoons 211 + Y_2Cu_2O_5$	p_4	~1070

in air and using a barium carbonate precursor, oxycarbonate phases are formed; one such phase is $Y_2Ba_3(CO_3)_nO_{6n}$, whose crystal structures are closely related to those of the parent oxide phases.

Further understanding of the formation of liquid phase may be obtained by considering the nature of the phase equilibria in the compatibility triangle: $123 + BaCuO_2 + CuO$. (Figure 8.12(a)). The binary system BaO–CuO (Figure 8.12(b)) shows a eutectic reaction (e_2): $L \rightleftharpoons BaCuO_2 + CuO$ occurring at ~920°C. The liquidus projection shows a eutectic valley extending into the ternary system to meet two other reaction curves, namely the eutectic: $L \rightleftharpoons BaCuO_2 + 211$ and the peritectic: $L + 211 \rightleftharpoons 123$, to give rise to a ternary invariant eutectic: $L \rightleftharpoons 123 + BaCuO_2 + CuO$, occurring at ~890°C. The vertical section (Figure 8.12(c)) which passes through the compositions of the 211 and 123 compounds includes the composition point m_1, which represents the maximum (saddle-point) on the peritectic curve: $L + 211 \rightleftharpoons 123$. Also the vertical section meets the binary system BaO–CuO at a composition close to that of the binary eutectic.

Processing of 123 Superconducting Material

A number of processing routes are available for the preparation of superconducting 123 in various forms and two of these are considered here in relation to the phase diagram and the microstructures of the products.

(i) The Preparation of Single Crystals from Melts

Reference to the liquidus projection (Figure 8.12(b)) shows that the 123 compound melts incongruently and cannot therefore be formed directly from liquid of stoichiometric composition. To produce crystals of 123 a powder mixture composition is chosen located in the primary 123 field of the liquidus i.e. the region bounded by the liquidus two-phase reaction curves: $p_3m_1p_1$, p_1e_1, e_1p_3. This region is small in its compositional range and the liquidus surface is steep, as illustrated by the substantial temperature difference between m_1 (1015°C) and e_2 (920°C). The vertical section (Figure 12c) also shows the steep liquidus boundary of the liquid + 123 region.

Consider, for example, a powder composition lying on the vertical section of Figure 8.12(c), and having a $YO_{1.5}$ content slightly less than that of point m_1, so that it is located in the primary 123 region of the liquidus. Assume that the mixture is heated to, and held at, a temperature above that of the eutectic e_1, say to ~900°C. Primary crystals of 123, shown by experiments to be of crystallographic shape, nucleate and grow in the liquid as equilibrium is approached; in principle, the proportion of primary crystals at equilibrium can be calculated from the phase diagram, if the details of the isothermal section are known. If the mixture of liquid plus 123 crystals is then slowly cooled then solidification proceeds via the relevant three-phase reaction curves until the eutectic composition, e_1 is reached. The solidified ceramic consists of the mixture of primary 123 crystals in a 'matrix' of crystals of $123 + BaCuO_2 + CuO$. The crystals of 123 can be mechanically separated from the solidified material.

(ii) Melt Textured Growth

This type of process aims to produce bulk artifacts of 123 compound with low angle grain boundaries, which provide the ability to carry high currents. The procedure adopted is to heat a powder mixture of 123 composition to a temperature above that of peritectic point m_1 (i.e. > 1015°C). The material is then held at this temperature so as to form crystals of 211 phase in liquid depleted in $YO_{1.5}$. At equilibrium, reference to the vertical section (Figure 8.12(c)) shows from the lever rule that the mole proportion of 211 phase is ~30%. The material is then slowly cooled (e.g. at <10°C per hour), sometimes with the application of a temperature gradient to aid directional growth. At temperatures below the m_1 peritectic, viz. in the 123 + liquid region of the phase diagram, elongated crystals of 123 nucleate and grow in the liquid. Experimental work has shown that nucleation of the 123 crystals does not occur preferentially at the 211/liquid interfaces, as might be expected, but independently in the liquid phase. The kinetics of growth of the crystalline plates of 123 depend on the solution of the 211 phase particles, inherited from the treatment at the higher temperature and the subsequent diffusion of Y to the solid 123/liquid interfaces. The 123 crystals grow as domains and consist of platelet crystals, of identical orientation, in which 211 particles are embedded; thus equilibrium is not reached, and there is evidence that the residual 211 particles exert a beneficial effect on the superconducting properties.

Problems

1. Assuming that the invariant eutectic e_1 (Table 1) occurs at a temperature of 890°C and at a composition (mol.%) of ~67 CuO; 1.5 $YO_{1.5}$; balance BaO, draw an isothermal section representing the 123 + $BaCuO_2$ + CuO region of the ternary section at a temperature 2°C above the ternary eutectic temperature.

2. Assuming that a uniform powder mixture consisting of (mol.%) 65 CuO; 10 $YO_{1.5}$; balance BaO, is heated to achieve equilibrium at 2°C above the ternary eutectic temperature as referred to in question 1 above, calculate, using the isothermal section, the proportion of liquid phase that will be present at equilibrium.

9. Answers to Problems

The use of triangular coordinate graph paper is recommended for questions involving the plotting of composition data. Alternatively, page 222 can be photocopied to provide graph paper.

CHAPTER 1

(a) Procedure: Plot the compositions of the three phases to form a tie-triangle and also plot the alloy composition. Measure the appropriate lengths from this diagram (e.g. cf. tx_1 and xx_1 in Figure 1.6) for the application of the 'centre of gravity' principle.

    Answer: ~55% Liquid, 10%α, 35%γ

(b) Procedure: From the lever rule the mid-point of the α–γ side of the tie-triangle is the composition required.

    Answer: ~48%A, 15%B, 37%C.

CHAPTER 3

Procedure: Draw possible forms of the binary diagrams assuming the liquidus and solidus curves to be smooth. Then, for diagrams (a) and (b) locate the binary alloy compositions corresponding to the quoted temperatures, e.g. for a liquidus of 950°C in binary system AB the alloy composition is ~70%B, 30%A, while the same liquidus in AC corresponds to an alloy composition of ~35%C, 65%A. On the ternary composition plot, draw a smooth curve, joining these two binary compositions, to represent the possible form of the liquidus isotherm (c.f. Figure 3.2). Draw a similar construction joining the solidus compositions for 950°C. The isothermal and vertical sections can then be constructed from diagrams (a) and (b) (see Figures 3.3 and 3.4).

    Answer: See Figures A1a and b.

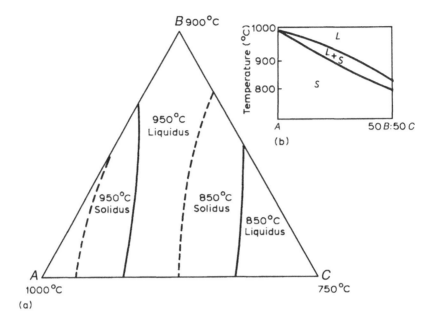

Fig. A1 (a)Liquidus and solidus projections showing isotherms for 950 and 850°C and (b) vertical section between *A* and the midpoint of *BC*.

CHAPTER 4

1. (a) (i) α/α + *L*; (ii) *L*/*L* + β; (iii) α + *L*/α + β + *L*; (iv) α + β + *L*/ α + β.
    (b) (i) *BEF*; (ii) *MEFN*.

2. Procedure: Use the given data to plot tie-triangles for 540 and 510°C, respectively, and calculate the proportions of the phases using the 'centre of gravity' principle.
    Answer: (a) ~36%α, 27%β. (b) Liquid at 540°C/Liquid at 510°C ~2.9.

3. (a) *L* → primary β, commencing at temperature T_l; *L* → α + β eutectic reaction, commencing at temperature T_e; see Figures 4.1 and 4.3. For the compositional changes involving *L* + α + β tie triangles: the eutectic reaction is completed at temperature T_s to give an α + β structure.
    (b) *MEFN*; *DEFG*.

4. Procedure: (a) From the data given (temperatures of the invariant reactions and the compositions of the phases involved) it follows that the binary invariant reactions are of the eutectic type *L* → α + β. Draw a smooth curve

joining the compositions of the eutectic liquid in systems *AB* and *AC*, to represent the liquidus projection. The solidus projection is similarly obtained by joining respectively the α compositions in *AB* and *AC* and the β compositions (cf Figure 4.4).

(b) The alloy containing 30%*A*, 55%*B* and 15%*C*, commences its solidification by the formation of the primary β phase. Solidification is completed by the eutectic reaction $L \rightarrow \alpha + \beta$ at a temperature when the α + β tie-line passes through the alloy composition point (see Figure 4.3). The position of this tie-line cannot be deduced accurately from the given data, but assuming a possible position for this line the lever-rule can be used to calculate the proportions of α and β.

Answer: (b) ~23%α, 77%β.

5.   (a) (i) $L + \alpha$; (ii) $L + \beta$; (iii) $L + \alpha + \beta$. (b) (i) *EDGF*; (ii) *AMN*.

6.   Procedure: (a) Plot the data in the form of three tie-triangles, representing the equilibria between liquid, α and β at 550, 520 and 500°C respectively. Apply the 'centre-of gravity' principle to the particular alloy quoted in relation to the 520°C tie-triangle.

Answer: (a) ~18%*L*, 45%α, 37%β.

(b) The data for the liquid, α and β corners of the tie-triangles enable parts of the liquidus and solidus projections to be represented, e.g. by drawing a curve through the liquid compositions for the equilibria at 550, 520 and 500°C respectively.

Draw a curve through the corners of the tie-triangles representing the liquid compositions and thus obtain a part of the liquidus projection. Proceed similarly for the α and β corners of the tie-triangles to obtain parts of the solidus projection. Then applying the 'tangent' procedure illustrated in Figure 4.17, it can be deduced that the three-phase reaction is a peritectic: $L + \beta \rightarrow \alpha$.

(c) From the fact that only three phases are present in the system, namely liquid, α and β, respectively, it follows that one of the binary systems shows complete solid solubility. Phase Rule considerations indicate that no invariant reactions occur within the ternary system, and if it is assumed that both systems *AB* and *BC* contain a peritectic reaction and system *AC* shows complete solid solubility then the liquidus projection will consist of a curve joining the peritectic liquid composition values in systems *AB* and *BC* (assume reasonable values for these compositions). The solidus projection will consist of two curves joining the α and β compositions involved in the binary peritectic reactions (assume reasonable values for these). The general form of the liquidus and solidus projections will be as in Figure 4.14.

To show an isothermal section in the solid state neglect the change of solid solubility with temperature, and the solidus projection may then be taken to correspond to the required isothermal section. Alternatively, assume that some decrease in the solubility of *A* and *C* in β, and of *B* in α occurs, and draw a possible isothermal section with boundaries displaced from the solidus projection. The isothermal section for 550°C will incorporate the 550°C tie-triangle composition data; Figure 4.15 provides a guide to the arrangement of the other regions in this section.

(d) Consideration of the alloy composition in relation to the liquidus projection and the solid state isothermal section shows that solidification commences with the separation of primary β, whereas the equilibrium structure of the solid is single phase α. Thus, the solidification sequence involves the peritectic reaction and the consumption of all the β phase during this reaction (see Figure 4.13). During the separation of primary β the liquid composition moves in a curved path on the liquidus surface, while the solid composition changes over the solidus surface; the precise paths cannot be deduced from the data given. At the end of the primary solidification the liquid + β tie-line forms one of the sides of a liquid + α + β tie-triangle. Consideration of the tie-triangle data shows that this stage is reached at a temperature close to 550°C. With falling temperature the $L + \alpha \rightarrow \beta$ reaction proceeds and the tie-triangles indicate its progress (cf Figures 4.12 and 4.13). At the end of the reaction the alloy composition point lies on the $L + \alpha$ side of a tie-triangle and from the tie-triangle data it can be seen that this stage is reached at ~520°C. The liquid composition now moves away, with falling temperature, from the peritectic liquid curve. As the composition moves over the α liquidus surface, α phase separates until the solidus is reached.

7. Procedure: Draw a line from the binary alloy composition (wt.%): 55Pb–45Zn parallel to the Zn–Sn side of the system. Determine the Sn contents corresponding to the intersections of this line with each of the tie-lines in the miscibility gap. Apply the lever-rule to each of the tie-lines to calculate the proportion of Pb-rich liquid.

    Answer: The proportion of Pb-rich liquid varies from ~60% in the binary Pb–Zn alloy to close to 100% in the alloy containing 55Pb 20Zn 25Sn.

CHAPTER 5

1· (a) (i) $L + \beta$; (ii) $L + \alpha + \gamma$. (b) (i) M_1DGM; (ii) CO_1OO_2.

2. (a) (i) M_1N_1NM, N_1DGN, M_1DGM, MNG; (ii) ADM_1 $ADGE$, AM_1MM_2, M_1MGD, M_2MGE, AEM_2. (b) (i) $L + \gamma/L + \beta + \gamma$; (ii) $L + \alpha + \beta/L + \alpha + \gamma/\alpha + \beta + \gamma$; (iii) $L + \alpha + \beta/L + \alpha + \gamma/L + \beta + \gamma$.

3. Procedure: (See Figure A2) (a) From the data, including negligible solid solubility and the absence of compounds, it can be deduced that the binary systems are of the eutectic type and that a ternary eutectic reaction $L \rightarrow A + B + C$ occurs at 600°C. The data for alloys 1, 2 and 3 show the compositions and temperatures of the binary eutectic reactions: $L \rightarrow A + B$; $L \rightarrow A + C$ and $L \rightarrow \beta + C$ respectively. To determine the ternary eutectic composition draw a line joining A to the composition of alloy 4;

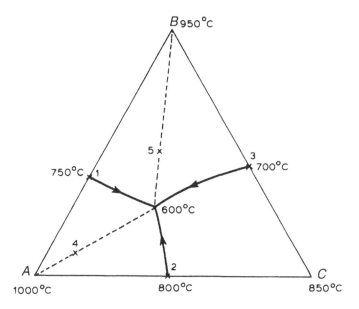

Fig. A2 Liquidus projection. Alloy compositions are marked X and numbered.

also draw a line joining B and the alloy 5 composition. Extend these lines to intersect one another; their point of intersection represents the ternary eutectic composition.

(b) The alloy composition lies on the line joining A to the ternary eutectic composition; apply the lever-rule to this line to calculate the proportion of primary A.

Answer: (a) See Figure A2. (b) ~65% of primary A.

4. (a) (i) A liquidus temperature of ~770°C may be deduced by 'linear interpolation between the 750 and 800°C isothermal curves. A better procedure is to extrapolate using a construction of part of a vertical section passing through Cu and the alloy composition point and extending as far as the intersection with the $L \rightarrow Cu + Cu_3P$ curve. Values of liquidus temperatures can be plotted corresponding to the intersections of this section with the isotherms; distances to represent the composition scale are measured from Figure 5.17 along the section from the Cu corner.
(ii)-(iv) Using the procedure described on pages 67-69: (ii) ~87% liquid; (iii) ~40% liquid, ~30%Cu, ~30%Cu$_3$P; (iv) ~13% primary Cu, ~65% binary eutectic (Cu + Cu$_3$P), ~22% ternary eutectic (Ag + Cu + Cu$_3$P). Note that the proportion of binary eutectic can be obtained by difference, after the primary Cu has been calculated from the appropriate tie-line and the ternary eutectic from the appropriate tie-triangle.
(b) See Figure A3.

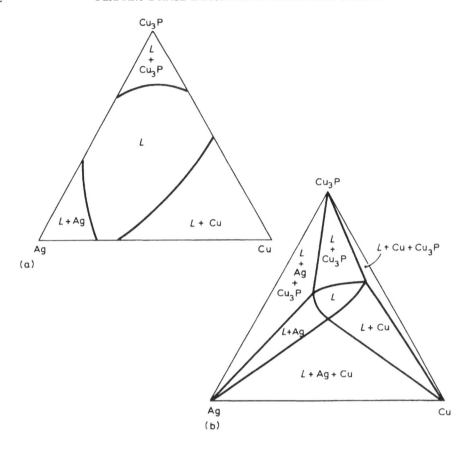

Fig. A3 Ag–Cu–Cu$_3$P system. Isothermal sections at (a) 800°C and (b) a temperature above the ternary eutectic but below the binary eutectics.

5. (a) See Figure A4. (b) System AB, $L + \beta \rightarrow \alpha$; BC, $L + \beta \rightarrow \gamma$; AC, $L \rightarrow \alpha + \gamma$; A BC, $L + \beta \rightarrow \alpha + \gamma$. (c) Alloy $\boxplus L \rightarrow \beta$; $L + \beta \rightarrow \alpha$; $L + \beta \rightarrow \alpha + \gamma$; $\alpha + \beta + \gamma$. By application of the 'centre of gravity' rule to the $\alpha + \beta + \gamma$ tie-triangle at 945°C: ~48%α, ~36%β, ~16%γ (d) $\oplus$ Alloy $+ L \rightarrow \beta$; $L + \beta \rightarrow \gamma$; β consumed; $L \rightarrow \gamma$; $L \rightarrow \alpha + \gamma$; $\alpha + \gamma$.

CHAPTER 6

1. (a) (i) and (ii) See Figure A5. Invariant reactions. System AB, $L \rightarrow \alpha + \beta$; BC, $L + \beta \rightarrow \delta$; $L \rightarrow \delta + \gamma$; AC, $L \rightarrow \alpha + \gamma$; ABC, $L + \beta \rightarrow \alpha + \delta$; $L \rightarrow \alpha + \delta + \gamma$.
(b) $\oplus L \rightarrow \beta$; $L + \beta \rightarrow \delta$; $L + \beta \rightarrow \alpha + \gamma$; liquid consumed; final structure $\alpha + \beta + \delta$. For a temperature of 651°C (i.e. just above the invariant reaction plane $L + \beta \rightarrow \alpha + \gamma$)

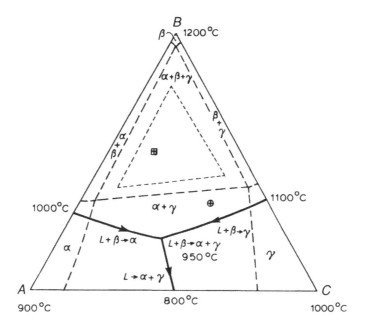

Fig. A4 —— Liquidus projection. – – – Isothermal section at room temperature. - - - - - Boundaries of α + β + γ region at 945°C.

apply the centre of gravity principle to the tie-triangle formed by joining the compositions of the peritectic liquid and the β and δ compositions. (Answer ~24%β). For 649°C use the α, β, δ tie-triangle. (Answer ~15%β).

2. (a) $L \rightarrow \alpha$; $L + \alpha \rightarrow \beta$ (along PX); $L \rightarrow \alpha + \beta + \delta$; final structure $\alpha + \beta + \delta$. (Note: mention is made on pages 69, 91 and 94 of the transition of a three-phase reaction from peritectic to eutectic to give rise to a ternary invariant eutectic reaction by intersection with two other eutectic curves. It is not necessary for the peritectic to show this transition in order to produce a ternary invariant eutectic; the reaction can remain peritectic in nature along the whole length of the liquid composition curve and this is the case for curve PX, representing $L + \alpha \rightarrow \beta$ in Figure 6.15).
    (b) $L \rightarrow \beta$; $L \rightarrow \beta + \delta$ (along MX); final structure $\beta + \delta$.

3. (a) (i) Liquidus ~570°C; $L \rightarrow Al$; $L \rightarrow Al + T$; $L \rightarrow Al + \beta + T$ at 450°C; final structure $Al + \beta + T$; (ii) Liquidus ~580°C; $L \rightarrow Al$, $L \rightarrow Al + S$; $L + Al \rightarrow S + T$ at 467°C; liquid consumed; final structure $Al + S + T$. (iii) Liquidus ~520°C; $L \rightarrow Al$; $L \rightarrow Al + S$; final structure $Al + S$. (iv) Liquidus ~635°C; $L \rightarrow Al$; $L \rightarrow Al + S$; solidus ~515°C; final structure $Al + S$. Note: in each of the above cases the primary phase (designated Al) is a solid solution of Cu and Mg in Al; the exact compositions

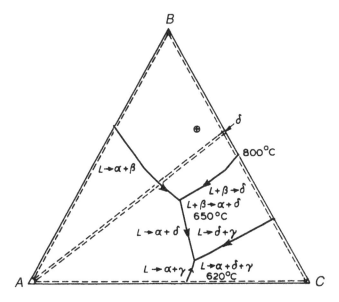

Fig. A5 ———— Liquidus projection. - - - Isothermal section at 600°C containing three-phase regions $\alpha + \beta + \delta$ and $\alpha + \delta + \gamma$ with the associated single-phase and two-phase regions.

of the solid solutions that form cannot be deduced from the data provided and hence the exact paths of the liquid composition changes during primary solidification are not known.

(b) For the binary Al–4·5%Cu alloy, the structure at 500°C is single phase Al solid solution, while for the ternary alloy it is Al solid solution + S phase. The solid solubility limit at 500°C is not given, but from the solidus projection it is seen that the Cu content of the solid solution in the ternary alloy is significantly lower than that in the binary alloy. On reheating the quenched alloys to 200°C, the structure of the binary alloy becomes Al solid solution + θ, while in the ternary alloy additional S phase will precipitate.

4. (a) See Figure A6a.

(b) Consider the intersections of the composition line with the isotherms on the liquidus surface (Figure 6.19b). From the SiO_2 end of the line the proportion of liquid increases linearly along the line up to 100% liquid, where the line intersects the 1100°C isotherm. Between this point and the intersection with the 1100°C isotherm in the primary $Li_2O \cdot SiO_2$ region, there is 100%, liquid at 1100°C (Figure A6b). For the remaining part of the composition line select various compositions; then obtain the proportions of liquid by applying the lever rule to the

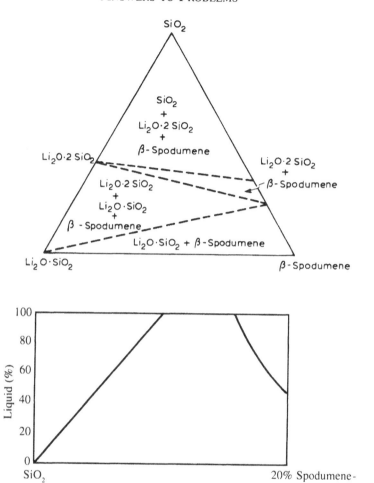

Fig. A6 (a) Isothermal section of the SiO_2-$Li_2O \cdot SiO_2$–$Li_2O \cdot Al_2O_3 \cdot 4SiO_2$ system at 970°C and (b) Variation of percentage liquid with composition at 1100°C along a line joining SiO_2 to 20% spodumene-80% $Li_2O \cdot SiO_2$.

tie-lines (for each point selected) drawn from $Li_2O \cdot SiO_2$ to the corresponding point on the 1100°C isotherm.

CHAPTER 8

Nb-Ta–C

See for guidance Figures 4.8, 4.16 and 6.12.

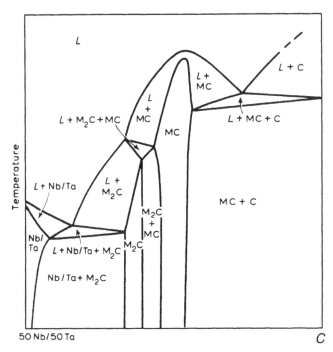

Fig. A7 Schematic vertical section of the Nb-Ta-C system from 50 Nb/50 Ta to C.

Answer: Figure A7 (schematic). The alloy lies in the M_2C + MC region; Incongruent melting of M_2C occurs by the reaction $M_2C \rightarrow L + MC$; melting continues with increase in temperature through the $L + MC$ region.

As–Ga–Zn

1. Procedure: From Figure 8.2d the source composition at 1000°C lies in the liquid + GaAs region of the system. Draw a tie-line from GaAs through the source composition to intersect the 1000°C isotherm shown on the liquidus projection; this intersection point represents the composition of the liquid in equilibrium with GaAs.

     Answer: ~51Ga 21As 28Zn in at.%.

2. (a) See Figure A8. Note: the solid solubility limits of As, GaAs, Zn_3As_2 and $ZnAs_2$ are very small; the lines joining respectively As-GaAs, As-$ZnAs_2$, GaAs-$ZnAs_2$, GaAs-Zn_3As_2 and Zn_3As_2-$ZnAs_2$ represent very narrow two-phase regions.
(b) Procedure: Convert the composition of the alloy and of the relevant solid phases (i.e. GaAs, Zn_3As_2, $ZnAs_2$) from atomic to weight percentages; plot

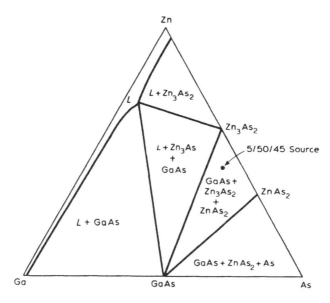

Fig. A8 Isothermal section of the Ga–As–Zn system at 700°C.

the resultant data as a tie-triangle and use the centre of gravity, principle.

Answer: ~11 GaAs, 54 Zn_3As_2 35 $ZnAs_2$ in wt.%.

3.   These sections are quasi-binary in nature and contain the equilibria $L \leftrightarrow$ GaAs + Zn_3As_2 and $L \leftrightarrow$ GaAs + $ZnAs_2$ respectively.

Au-Pb-Sn

1. $L \to AuSn_2$; $L \to AuSn_2$ + Pb; $L + AuSn_2 \to$ Pb + $AuSn_4$; $L \to AuSn_4$ + Pb; $L \to AuSn_4$ + Pb + Sn. ~ 40% Pb, 26% Sn, 34% $AuSn_4$.

2.   Procedure : In Figure 8.3b draw a line from Au to the solder composition; the amount of Au that can dissolve at 280°C is given by the composition corresponding to the intersection of this line with the 280°C isotherm.

Answer: ~ 26% Au.

3.   Using the procedure indicated from question 2 above, the 40Pb-60Sn solder is found to dissolve ~10%Au at 220°C and the 50Pb–50Sn solder ~12% Au at 240°C. Considering the solidification sequences of the liquids formed by solution of Au for these two cases respectively, both will deposit primary $AuSn_2$, and the liquid compositions will then reach P_2 via the reaction $L \to$ Pb + $AuSn_2$. At P_2, $L + AuSn_2$

$\rightarrow$ Pb + AuSn$_4$ and under equilibrium AuSn$_2$ is consumed, since the overall compositions lie in the L + Pb + AuSn$_4$ region. Applying the centre of gravity construction the proportion of liquid (composition P$_2$) is ~80% for the 40Pb–60Sn solder as compared with ~45% for the 50Pb-50Sn solder. Solidification is completed by the sequence $L \rightarrow$ Pb + AuSn$_4$; $L \rightarrow$ Pb + Sn + AuSn$_4$, (at E). Application of the centre of gravity principle to the Pb–Sn–AuSn$_4$ triangle shows that ~34% AuSn$_4$ is formed in the 40Pb-60Sn solder as compared with ~37% AuSn$_4$ in the 50Pb–50Sn solder.

Ni-Al-Cr

1. (a) Consider alloy X (Figure 8.4d): $L \rightarrow$ primary β; the liquid composition moves on to the $L + \beta \rightarrow \gamma'$ peritectic curve; as the peritectic reaction proceeds the alloy composition lies in the three-phase $L + \beta + \gamma'$ region; at 1340°C the liquid composition reaches L_p and the invariant reaction occurs $L_p + \gamma'_p \rightarrow \beta_p + \gamma_p$; the liquid is consumed since X lies in the $\beta_p + \gamma_p + \gamma'_p$ tie-triangle and solidification is thus completed.
    (b) Alloy containing 75 at.%Ni, 22 at.%Al, 3 at.% Cr (represented schematically by Y in Figure 8.4d); $L \rightarrow$ primary γ; the liquid composition moves on to the $L \rightarrow \gamma + \gamma'$ eutectic curve; solidification is completed to form $\gamma + \gamma'$.

2. Schematic representation of the transition from β - γ to α - γ' equilibrium at 1000°C (a) Above invariant reaction temperature, (b) at invariant reaction, and (c) below invariant reaction temperature. (See Figure A9).

3. Figures 8.4e and f show that the solubility of Al in γ decreases with increase in Cr content; similarly the $\gamma'/\gamma + \gamma'$ boundary moves to lower Al content with Cr additions. Considering a series of alloys with 15 at.%Al and increasing Cr content up to 20%, at 750°C (Figure 8.4f) the proportion of γ' increases, as indicated by the tie-line data. The lattice parameter mismatch between γ and γ' is small, and varies with composition, becoming zero for certain values of γ and γ' compositions. Considering a series of alloys with 75 at.% Ni (Figure 8.4g), the $\gamma/\gamma + \gamma'$ transus temperature decreases with increase in Cr content; also in the $\gamma + \gamma'$ region at a given ageing temperature (e.g. 750°C), the proportion of γ' decreases (note that the γ/γ' tie-lines lie quite close to the plane of this vertical section at 750°C).

Ni–Al–Ti

(a)

p_3	$L +$	β_1	$\rightarrow$	γ'		1395 °C
e_1	$L \rightarrow$	γ	+	γ'		1385 °C
e_2	$L \rightarrow$	γ	+	η		1304 °C
U_1	$L +$	γ	$\rightarrow$	γ'	$+ \eta$	1302 °C
U_2	$L +$	β_1	$\rightarrow$	γ'	$+ H$	1287°C

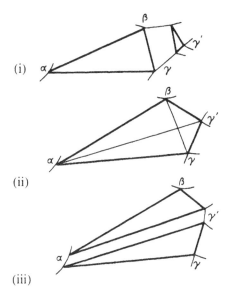

(i)

(ii)

(iii)

Fig. A9 Equilibrium associated with the invariant reaction involving β, γ, α and γ'.

E_1 $L \rightarrow$ γ' + η + H 1277°C
e_3 $L \rightarrow$ η + $β_2$ 1124°C
U_3 $L +$ η $\rightarrow$ $β_2$ + H 1107°C

(b) $L \rightarrow β_1; L \rightarrow β_1 + H; L + β_1 \rightarrow H + γ'$; liquid consumed; structure at 1150°C, $H + β_1 + γ'$.

Fe–Cr–C

1. Ternary invariant reactions:
    1585°C: $L + M_3C_2$ $\rightarrow$ graphite + M_7C_3
    1415°C: $L + M_{23}C_6$ $\rightarrow$ M_7C_3 + α
    1275°C: $L + α$ $\rightarrow$ γ + M_7C_3
    1230°C: $L +$ graphite + M_7C_3 $\rightarrow$ M_3C (see Figure 5.13)
    1160°C: $L + M_7C_3$ $\rightarrow$ γ + M_3C
    1155°C: $L + M_3C$ $\rightarrow$ γ + graphite

2. $L \rightarrow$ primary α (commencing at ~1510°C); liquid composition changes until it reaches the $L + α \rightarrow γ$ curve; the liquid is consumed during this reaction and the resulting structure on cooling to just below the solidus(~1430°C) is α + γ.

3. At room temperature, under equilibrium, the alloy consists of ferrite + $M_{23}C_6$. On heating to 1200°C a fully austenitic structure is produced; rapid quenching leads to martensite formation (together with a small amount of retained austenite); tempering

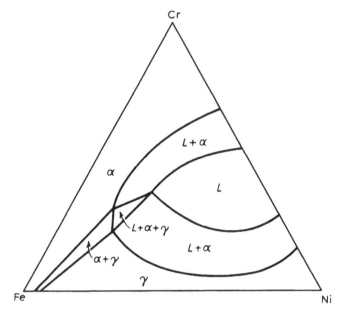

Fig. A10 Isothermal section of the Fe–Cr–Ni system at 1400°C.

at 700°C produces a structure of ferrite + $M_{23}C_6$. (The chromium content of the ferrite is reduced to less than 9% by the carbide precipitation and this adversely affects the corrosion resistance.) At 850°C the alloy composition lies in the $\gamma + M_{23}C_6$ phase region; the γ phase contains ~9%Cr, 0·2%C, and rapid quenching produces martensite which on tempering transforms to ferrite + $M_{23}C_6$.

(Note: the lever rule cannot be applied directly to Figures 8.6b and c since triangular coordinates are not used and the percentage carbon scale is expanded relative to that for chromium).

Fe-Cr-Ni

1. See Figure A10.

2. $L \to \alpha$ (commencing at ~1430°C); liquid composition moves on to the $L \to \alpha + \gamma$ reaction curve. The alloy composition lies close to the $\alpha + \gamma$ side of the $L + \alpha + \gamma$ tie-triangle in the 1400°C isothermal section. Thus the solidus for the alloy is ~1400°C; the application of the lever rule to the α–γ tie-line gives the proportions of α and γ as ~75 and 25%, respectively.

3. The Fe–18Cr–12Ni composition lies very close to the L, α, γ reaction curve; solidification commences at ~1470°C and proceeds via the $L \to \alpha + \gamma$ reaction to produce an $\alpha + \gamma$ structure at ~1420°C. On cooling in the solid state the structure

becomes single phase γ at a temperature not far below the solidus.[46] On further cooling, under equilibrium, σ phase begins to precipitate from γ at just above 650°C (Figure 8.7d). If the alloy is rapidly cooled from 1000°C the austenite structure is retained to room temperature.

4. From Figure 8.7d, along the line of constant Ni content (5%) alloys containing between ~21 and 30%Cr lie in the α + γ + σ tie-triangle and the proportion of σ increases from 0 to ~50% (calculated using the centre of gravity construction). From ~30 to 45% Cr in the γ + σ region the proportion of δ increases to 100% (calculated from the lever rule and estimated positions of tie-lines). The single phase region is traversed between ~45 and 53% Cr and then as the Cr increases to ~63% (i.e. across the σ + α' region), the σ -phase content diminishes to zero.

5. Application of the centre-of-gravity procedure to the α + γ + σ tie-triangle shows that the %σ is ~10, in good agreement with the calculated value.

Fe–O–Si

1. Applying the lever rule along the part of the fayalite–magnetite line between the position of the 1400°C isotherm and the composition of magnetite, ~0.45 tonnes of solid magnetite is present.

2. Table showing compositions (wt. fraction) of the phases coexisting in equilibria at temperatures in the range 1350-1150°C in any alloy of composition (wt. fraction) 0.720 FeO, 0.080 SiO_2, 0.200 Fe_2O_3.

	Liquid			Wüstite		
	FeO	SiO_2	Fe_2O_3	FeO	SiO_2	Fe_2O_3
D. Start of solidification, 1326.4°C	0.7200	0.0801	0.1999	0.7286	0.0000	0.2714
F. Wüstite at start of eutectic reaction, 1167°C				0.7447	0.0000	0.2553
E. Liquid at start of eutectic reaction, 1167°C	0.6739	0.2250	0.1010			
B. Wüstite at end of eutectic reaction, 1164.8°C				0.7258	0.0000	0.2742
C. Liquid at end of eutectic reaction, 1164.8°C	0.6673	0.2245	0.1083			

Note: The composition of the fayalite formed in the eutectic reaction L --→ wüstite + faylite is: FeO, 0.7051; SiO_2 0.2949. The slope of the eutectic curve is very shallow and the eutectic reaction occurs over a very narrow temperature range. The thermodynamically calculated value for point B is richer in Fe_2O_3 relative to point D in the schematic diagram of Fig. 8.6 d.

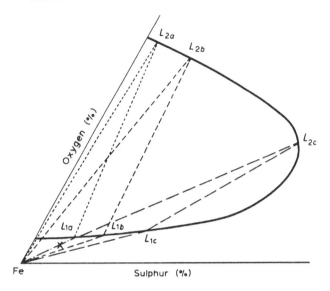

Fig. A11 Schematic representation of part of the liquidus projection of the Fe–O–S system.

Fe-O-S

1. See Figure A11. Three tie-triangles (FeL_1L_2) are shown schematically in relation to the liquid miscibility gap to represent the progress of the monotectic reaction in alloy X. The three temperatures illustrated are indicated by subscripts a, b, c (see also reference 2).

   .... represents the beginning of the monotectic reaction (X lies on the Fe L_{1a} side of the Fe $L_{1a} L_{2a}$ tie-triangle).

   - - - - - at a lower temperature X lies within the FeL_{1b} L_{2b} tie-triangle.

   – – – – the end of the reaction is reached when X lies on the FeL_{2c} side of the FeL_{1c} L_{2c} tie-triangle.

2. For alloys that undergo the monotectic reaction the weight fraction of inclusions formed increases as the oxygen and sulphur contents of the alloy increase. However, the distribution of the compositions of the inclusions depends only on the oxygen/sulphur ratio. For alloys of high oxygen/sulphur ratio the bulk of the inclusions are rich in oxygen, while for low oxygen/sulphur ratios the majority of the inclusions are sulphur rich.

Ag–Pb–Zn

1. See Figure A12.

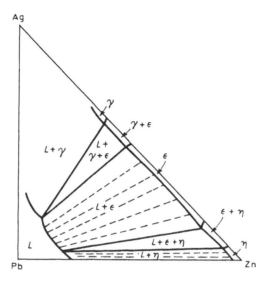

Fig. A12 Schematic isothermal section of Ag–Pb–Zn system at ~330°C.

2. The eutectic reaction $L \rightarrow Pb + \eta$ runs from the temperature of the invariant down to the Pb–Zn binary eutectic at 318°C (Figure 8.9e). Thus liquid solidifying along this curve is progressively depleted in silver content. By repeated zinc additions, the silver content of a bullion can be reduced to any desired figure as the Pb–Zn eutectic is approached. If the ternary invariant involving L, Pb, ε and η were a eutectic then the ternary eutectic composition would represent the lowest silver content attainable in lead by the use of zinc additions.

Al–Fe–Si (see Figs. 8.9 e and f)

   $L \rightarrow Al$; $L \rightarrow Al + Al_3Fe$; $L + Al_3Fe \rightarrow Al + \alpha AlFeSi$; Al_3Fe consumed; $L \rightarrow Al + \alpha AlFeSi$; $Al + \alpha AlFeSi$.

Al–Fe–Mn (see Figs. 8.9 g, h and i)

   654°C: $L \rightarrow Al + Al_3Fe + Al_6Mn$
   730°C: $L + Al_3Fe \rightarrow Al_4Mn + Al_6Mn$

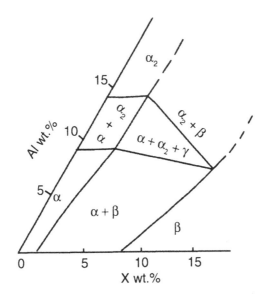

Fig. A13 Partial isothermal section of the Ti-Al-X system at 700°C.

Al–Mn–Si (see Figs 8.9 j and k)

> 690°C: $L + Al_4Mn \rightarrow Al_6Mn + \beta AlMnSi$
> 655°C: $L + \beta AlMnSi \rightarrow Al_6Mn + \alpha AlMnSi$
> 649°C: $L + Al_6Mn \rightarrow Al + \alpha AlMnSi$

Ti–Al Based Alloys (see Figs A13 and A14)

Al_2O_3–MgO–SiO_2

1. The section can be drawn from the information shown in Figure 8.10d (see Figure A15a).
2. (a) $L \rightarrow$ spinel (Al_2O_3-rich); $L \rightarrow$ spinel + corundum; L + corundum $\rightarrow$ spinel + mullite at 1578°C; the overall composition lies just within the L + spinel + mullite region at 1578°C, thus corundum is consumed; $L \rightarrow$ spinel + mullite.
   (b) $L \rightarrow$ spinel (Al_2O_3-rich); subsequent liquid path depends in detail on the spinel composition but liquid undergoes the L + spinel + mullite $\rightarrow$ sapphirine reaction at

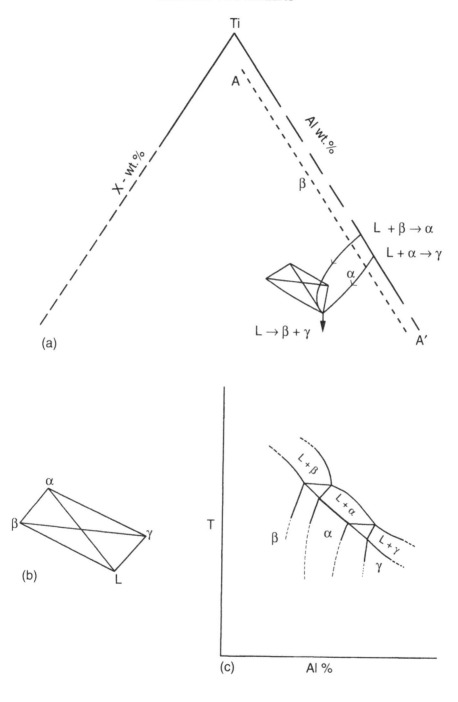

Fig. A14 (a) Ti–Al–X system. Schematic part of liquidus projection showing invariant reaction $L + \alpha \rightarrow \beta + \gamma$. (b) Enlarged view of the tie-triangles in Figure A14a merging at the invariant plane. Triangles above the plane $L + \beta + \alpha$: $L + \alpha + \gamma$. Triangles below the plane $\beta + \alpha + \gamma$; $L + \beta + \gamma$ and (c) Schematic vertical section along AA'.

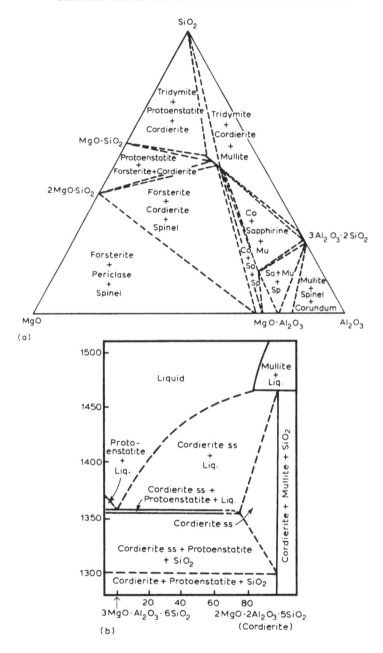

Fig. A15 Al$_2$O$_3$-MgO-SiO$_2$ system. (a) Isothermal section representing the solid state constitution (at a temperature of ~1350°C) derived from the data in Figure 7.10d.[56] (Reproduced from W. Schreyer and J. F. Schairer, *J. of Petrology*, 1961, **2**, **3**, p.349, with permission from Oxford University Press, Oxford) and (b) Vertical section between cordierite and the composition 3MgO·Al$_2$O$_3$·6SiO$_2$ (which lies close to the composition of the eutectic: liquid tridymite + cordierite + protoenstatite).[81] (Reproduced from W. Schreyer and J. F. Schairer, *J. of Petrology*, 1961, **2**, **3**, p.352, with permission from Oxford University Press, Oxford).

1482°C; mullite is consumed and the remaining liquid solidifies along the L + spinel $\rightarrow$ sapphirine curve running down from 1482°C.

(c) $L \rightarrow$ spinel (Al$_2$O$_3$-rich); $L \rightarrow$ spinel + corundum at 1578°C; L + corundum $\rightarrow$ spinel + mullite; corundum is consumed; $L \rightarrow$ spinel + mullite; at 1482°C, L + spinel + mullite $\rightarrow$ sapphirine; L, spinel and mullite consumed, leaving only sapphirine.

(d) $L \rightarrow$ cristobalite/tridymite; $L \rightarrow$ tridymite + cordierite.

(Note: when spinel is the primary phase, its precise composition is not known because of the wide solid solubility range and hence the path followed by the liquid composition over the primary liquidus surface is not known).

3. See procedure discussed on pages 68 and 69, Figure 5.5.

(a) Melting commences at the 1365°C eutectic (forsterite + protoenstatite + cordierite $\rightarrow$ liquid). The ceramic composition lies on the tie-line joining forsterite to the ternary eutectic composition. Application of the lever rule to this tie-line shows that ~20% liquid forms by melting of the ternary eutectic. On continued heating the proportions of liquid present at specific temperatures may be calculated by application of the lever rule to the forsterite-liquid tie-lines (the liquid compositions being located at points on the liquidus isotherms). Thus ~30% liquid is present at 1600°C and ~39% at 1700°C; melting is completed at ~1850°C.

(b) Melting commences at the 1355°C eutectic (protoenstatite + tridymite + cordierite $\rightarrow$ liquid). Application of the centre of gravity principle to the liquid + protoenstatite + tridymite tie-triangle shows that ~ 30% liquid forms. On continued heating, melting proceeds with the liquid composition changing along the liquid/protoenstatite/tridymite reaction curve. (The ceramic composition lies on this curve). At 1400°C ~35% liquid is present, and at ~1500°C melting is completed.

(c) Melting commences at the 1355°C eutectic and ~40% liquid forms, assuming the maximum solubility range for cordierite shown in Figures 8.10d and A15a. However, it should be noted that the range is not known accurately, and a smaller range is shown in the vertical section of Figure A15b[81], (the alloy composition lies close to this vertical section). Melting is complete at ~1430°C.

The variation of the proportion of liquid with temperature is important in the firing behaviour of ceramic compositions where the formation of ~ 25-40% of viscous liquid phase acts as a bond and aids densification. In the forsterite composition, (a), the amount of liquid does not increase rapidly with temperature (i.e. steep liquidus) and a broad firing range is therefore available. The shallower liquidus profiles of the steatite and cordierite type materials, (b) and (c), are not so favourable, but the addition of other materials may be used to improve the characteristics. It should be noted that in practice the processing of these ceramics from this system involves the use of clay and talc as raw materials (for a discussion of associated structural changes and of the kinetics of firing of ceramics see reference 79). It should also be borne in mind that the liquid phase commonly forms a glass on cooling instead of crystallizing.

3. The 42% Al$_2$O$_3$-58% SiO$_2$ material is solid at 1460°C but with 1% MgO the composition lies in the liquid (g)+ mullite + cristobalite region (Figure 8.10e).

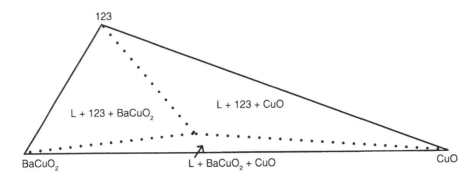

Fig. A16 Isothermal section representing the equilibria involving liquid, $BaCuO_2$, 123 and CuO, associated with the ternary eutectic reaction.

Applying the centre of gravity principle to the tie-triangle the percentage liquid is ~10%.

$BaO–CuO–YO_{1.5}$

1. (see Figure A16).
2. The isothermal section(see Figure A16) consists essentially of three tie-triangles containing respectively: $L + 123 + BaCuO_2$; $L + BaCuO_2 + CuO$; and $L + 123 + CuO$; since the temperature is only 2°C above the ternary eutectic temperature, the compositions of the liquid will virtually coincide with the ternary eutectic composition, and the two-phase regions in the section will be so narrow that they are, in effect, represented by lines. The alloy in question lies in the $L + 123 + CuO$ triangle, and application of the centre-of -gravity rule gives the proportion of liquid as ~ 15%.

10. Reviews and Compendiums of Experimental Phase Diagram Determination, Current Activities in Phase Diagram Determination and Information Regarding Software Packages for the Thermodynamic Calculation of Phase Equilibria

The activity in the experimental determination of phase diagrams probably reached its maximum effort during the middle of the century and some of the major compendiums of binary and ternary diagrams were published following this time. For example, the three volumes of the series *Constitution of Binary Alloys* were published,[1-3] which still act as a prime reference source for modern use. In the field of ceramics, a seminal work *Phase Diagrams for Ceramists* was published[4] in 1956 and a major effort has since been maintained by the American Ceramics Society with the publication of numerous supplements.[5] Further compendiums of binary phase diagrams were published by Mofatt,[6] Massalski et al.,[7, 8] Predel[9] and Okamoto,[10] and a number of publications concentrating on specific metals based binary alloys have been released.[11-24]

At the same time, major publications with respect to ternary and higher order systems were being written. For example, the book *Phase Diagrams for Ceramists* included many multi-component diagrams[4] and phase equilibria in oxides for steel making gave valuable insight into slags.[25] In the field of alloys, Haughton and Prince published the book, *Constitutional Diagrams of Alloys,*[26] and Prince followed this with bibliographies of cited literature involving phase equilibria in ternary and higher order alloys.[27, 28] Interestingly, major effort had already been undertaken in Al alloys and work that still acts as a standard in the modern day was published.[29-31]

More recently detailed reviews and assessments of ternary and higher order diagrams have been published for Ag, Al and Au alloys,[32-34] Cu-based alloys,[35] Fe-based alloys,[36-38] Ni-based alloys,[39] and Metal-Boron-Carbon,[40] BN and SiN based systems.[41] A comprehensive review of published work on ternaries, involving mainly metallic systems, has recently been released[42] while the American Ceramic Society continues to publish reviews of work done on ternary and higher order ceramic systems.[5] Note should also be made of the *Red Books* series that provides yearly summaries of published phase diagrams and constitutional work for metallic systems.[43] While most work on phase diagrams has centred on metals and high temperature ceramics, other areas have found substantial use for phase diagrams, for example, high T_c superconductors[44] and polymers.[45]

Experimental determination of phase diagrams has continued around the world, with new methods being developed that permit the quite rapid determination of phase equilibria to be achieved. To help co-ordinate this effort and ensure that duplication of experimental work, the International Programme for Alloy Phase Diagram Data was set up. This effort is co-ordinated by the Alloy Phase Diagram International Commission (APDIC) whose members currently include:

Argentine Committee for Phase Diagrams, Argentina,
ASM International, U.S.A.,
Baikov Institute of Metallurgy, Russia,
Chimica Generale, University of Geneva, Italy,
Chinese Phase Diagram Committee, PR China,
Deutsche Gesellschaft für Materialkunde, Germany,
Comite Brasileiro de Diagramas de Fases de Materiais, Brazil,
Deutsche Gesellschaft für Materialkunde, Germany,
Groupe de Thermodynamique et Diagrammes de Phases, France,
The Indian Institute of Metals, India,
The Institute of Materials, U.K.,
Japanese Committee for Alloy Phase Diagrams, Japan,
Materials Science International Services, Germany,
Max-Planck-Institut für Metallforschung, Germany,
Polish Phase Diagram Committee, Poland,
The Royal Institute of Technology, Sweden and
Ukrainian Phase Diagram and Thermodynamic Commission, Ukraine.

Many of these members either publish Journals for the specific purpose of reporting new experimental work or release compendiums of assessments of experimentally determined binary and ternary phase diagrams. There are a number of Journals that provide an excellent source for published work on phase diagram determination[46-57] and there has been a series of conferences, loosely grouped to together, concentrating on the use of phase diagrams.[58-61]

Alongside this activity in experimental measurement, a new field came to the fore during the latter part of the 20th century. This is based on the thermodynamic calculation of phase diagrams, often known under the acronym CALPHAD,[62] and has been described in Chapter 2. Presently, there are numerous software programmes which can be used for calculating phase diagrams or phase equilibria in multi-component alloys and ceramics.

Historically, the first software programmes dedicated to modelling of alloy phase equilibria and made available for general use originated from MANLABS[63] and the National Physical Laboratory.[64] Shortly after this, further programmes became available, for example F*A*C*T[65] (which later incorporated SolGasMix[66]), Thermo-Calc[67] and MTDATA.[68] An interesting development of new computer technology, the ability to use programmes on-line, was utilised by a number of the early programmes, including THERMODATA.[69] These formed the established set of programmes with the capability for calculating binary and ternary phase diagrams as reported by Bale and Eriksson[70] in 1990. Mention should also be made of the programmes BINGSS and TERGSS,[71] part of a suite of software used for least-square optimisation of thermodynamic parameters, whose source code has been made freely available to interested parties on request. These programmes have since been extended to include the facility for calculation of multi-component alloys.[72] Since then other programmes have been released for phase equilibria calculations, such as ChemSage,[73] PANDAT[74] and EQUILIB.[75]

The programmes mentioned above are not meant to provide an unequivocal list of current programmes, as the authors may be unaware of new developments at the time of writing the book. They also do not include packages specifically aimed at the calculation of complex chemical reactions. The main aim has been to introduce the reader to the variety of software that is now available, with suitable scientific references that can be followed through.

10.1 SOURCE REFERENCES

1. M. Hansen and K. Anderko, *Constitution of Binary Alloys*, McGraw-Hill, 1958.
2. R.P. Elliott, *Constitution of Binary Alloys*, 1st Supplement, McGraw-Hill, 1965.
3. F.A. Shunk, *Constitution of Binary Alloys*, 2nd Supplement, McGraw-Hill, 1969.
4. E.M. Levin, H.F. McMurdie and F.P. Hall, *Phase Diagrams for Ceramists*, The American Ceramic Society, **1**, 1956, and supplements.
5. *Phase Equilibria Diagram*, The American Ceramic Society, **1-13**, 1964-2001.
6. W.G. Moffatt, *The Handbook of Binary Phase Diagrams*, Genium Publishing, 1986.
7. T.B. Massalski, J.L. Murray, L.H. Bennett and H. Baker, eds., *Binary Alloy Phase Diagrams*, ASM International, U.S.A., 1986.
8. T.B. Massalski, H. Okamoto, P.R. Subramanian and L. Kacprzak eds., *Binary Alloy Phase Diagram*, 2nd Edition, ASM International, U.S.A., 1990.
9. B. Predel, *Phase Equilibria, Crystallographic and Thermodynamic Data of Binary Alloys: Landolt-Börnstein*, New Series, O. Madelung, ed., Springer, **5**, 1991-1998.
   Subvols. (a) Ac–Au to Au–Zr, (b) B–Ba to C–Zr, (c) Ca–Cd to Co–Zr, (d) Cr–Cs to Cu–Zr, (e) Dy–Er to Fr–Mo, (f) Ga–Gd to Hf–Zr, (g) Hg–Ho to La–Zr, (h) Li–Mg to Nd–Zr, (i) Ni–Np to Pt–Zr and (j) Pu–Re to Zn–Zr.
10. H. Okamoto, *Desk Handbook: Phase Diagrams for Binary Alloys*, ASM International, U.S.A., 2000.
11. M.E. Kassner and D.E. Peterson eds., *Phase Diagrams of Binary Actinide Alloys*, ASM International, U.S.A., 1995.

12.   L. Tanner and H. Okamoto, eds., *Phase Diagrams of Binary Beryllium Alloys*, ASM International, U.S.A., 1987.

13.   P.R. Subramanian, D.J. Chakrabati and D.E. Laughlin, eds., *Phase Diagrams of Binary Copper Alloys*, ASM International, U.S.A., 1994.

14.   H. Okamoto and T.B. Massalski, eds., *Phase Diagrams of Binary Gold Alloys*, ASM International, U.S.A., 1987.

15.   F.D. Manchester, ed., *Phase Diagrams of Binary Hydrogen Alloys*, ASM International, U.S.A., 2000.

16.   C.E.T. White and H. Okamoto, eds., *Phase Diagrams of Indium Alloys and Their Technical Applications*, ASM International, U.S.A., 1991.

17.   O. Kubaschewski, *Iron-Binary Phase Diagrams*, Springer, 1982.

18.   H. Okamoto, ed., *Phase Diagrams of Binary Iron Alloys*, ASM International, U.S.A., 1993.

19.   A.A. Nayeb-Hashemi and J.B. Clark, eds., *Phase Diagrams of Binary Magnesium Alloys*, ASM International, U.S.A., 1988.

20.   P. Nash, ed., *Phase Diagrams of Binary Nickel Alloys*, ASM International, U.S.A., 1991.

21.   S.P. Garg, M. Venkatraman, N. Krishnamurthy and R. Krishnan, eds., *Phase Diagrams of Binary Tantalum Alloys*, Indian Institute of Metals, India, 1996.

22.   J.L. Murray, ed., *Phase Diagrams of Binary Titanium Alloys*, ASM International, U.S.A., 1987.

23.   S.V. Naidu and P.R. Rao, eds., *Phase Diagrams of Binary Tungsten Alloys*, Indian Institute of Metals, India, 1991.

24.   J.F. Smith, ed., *Phase Diagrams of Binary Vanadium Alloys*, ASM International, U.S.A., 1989.

25.   A. Muan and E.F. Osborn, *Phase Equilibria Among Oxides in Steel Making*, Addison-Wesley, 1965.

26.   J.L. Haughton and A. Prince, *The Constitutional Diagrams of Alloys*, The Institute of Metals, U.K., 1956.

27.   A. Prince, *Multicomponent Alloy Constitution Bibliography, 1955-1973*, The Metals Society, London, U.K., 1978.

28.   A. Prince, *Multicomponent Alloy Constitution Bibliography, 1974-1977*, The Metals Society, London, 1981.

29.   G. Phragmen, *Journal of Institute of Metals*, **77**, 1950, 489-522.

30(a). H.W.L. Phillips, *Annotated Equilibrium Diagrams of Some Aluminium Alloy Systems*, The Institute of Metals, U.K., 1959.

30(b). H.W. Phillips, *Equilibrium Diagrams of Aluminium Systems*, The Aluminium Development Association, 1961.

31.   G. Petzow, G. Effenberg and F. Aldinger, eds., *Ternary Alloys, A Comprehensive Compendium of Evaluated Constitutional Data*, VCH, **1-3**, 1998-1994.
     Vol.1. Ag–Al–Au to Ag–Cu–P, 2. Ag–Cu–Pb to Ag–Zn–Zr, 3. Al–Ar–O to Al–Ca–Zn, 4. Al–Cd–Ce to Al–Cu–Ru, 5. Al Cu–S to Al–Gd–Sn, 6. Al–Gd–Tb to Al–Mg–Sc, 7. Al–Mg–Se to Al–Ni–Ta, 8. Al–Ni–Tb to Al–Zn–Zr, 9. As–

Au–Ca to As–Co–V, 10. As–Cr–Fe to As–I–Zn, 11. As–In–Ir to As–Yb–Zn, 12. Au–B–Co to Au–Ge–La, 13. Au–Ge–Li to Au–Tl–Zn, 14. Ag–Al–Li to Ge–Li–Nd, 15. Ge–Li–Ni to Li–V–Zr.

32.   L.F. Mondolfo, *Aluminium Alloys: Structure and Properties,* Butterworth, London, U.K., 1976.

33.   G. Effenberg, ed., *COST 507, Definition of Thermochemical and Thermophysical Properties to Provide a Database for the Development of New Light Alloys, Critical Evaluation of Ternary Systems,* Office for Official Publications of the European Communities, **3**, 1998.

34.   A. Prince, G.V. Raynor and D.S. Evans, *Phase Diagrams for Ternary Gold Alloys,* The Institute of Metals, U.K., 1986.

35.   Y.A. Chang, J.P. Neumann, A. Mikula and D. Goldberg, *Phase Diagrams and Thermodynamic Properties for Ternary Copper-Metal Systems,* International Copper Research Association, and Y.A. Chang, J.P. Neumann and U.V. Choudary, *Phase Diagrams and Thermodynamic Properties of Ternary Copper-Sulphur-Metal Systems,* International Copper Research Association, 1979.

36.   V. Raghavan, *Phase Diagrams of Ternary Iron Alloys,* (Part 1), 1987. *Ternary Systems Containing Iron and Sulphur,* (Part 2), 1988. *Ternary Systems Containing Iron and Phosphorous,* (Part 3), 1988. *Ternary Systems Containing Iron and Oxygen,* (Part 5), 1989. *Fe-X1-X2 Systems,* **1** and **2**(Part 6), 1992. (Part 1), American Society for Metals, (Parts 2, 3, 5 and 6), The Indian Institute of Metals, India.

37.   V. Raghavan, *Phase Diagrams of Quaternary Iron Alloys,* The Indian Institute of Metals, India, 1996.

38.   V.G. Rivlin and GV. Raynor, *Phase Equilibria in Iron Ternary Alloys,* The Institute of Metals, U.K., 1988., Originally Published in *International Metals Reviews.*

39.   K. Gupta, *Phase Diagrams of Ternary Nickel Alloys,* Indian Institute of Metals, India, (Parts 1and 2).

40.   P. Rogl, *Phase Diagrams of Ternary Metal-Boron-Carbon Systems,* ASM International, U.S.A., 1998.

41.   P. Rogl and J.C. Schuster, *Phase Diagrams of Ternary Boron Nitride and Silicon Nitride Systems,* International Copper Research Association, ASM International, U.S.A., 1992.

42.   P. Villars, A. Prince and H. Okamato, *Handbook of Ternary Alloy Phase Diagrams,* ASM International, U.S.A., 1995.

43.   G. Effenberg, ed., *The Red Books,* MSI and VINITI.

44.   J.D. Whitler and R.S. Roth, *Phase Diagrams for High T_c Superconductors,* The American Ceramic Society, U.S.A., 1991.

45.   A.H. Windle, *A Metallurgist's Guide to Polymers, in Physical Metallurgy,* R.W. Cahn and P. Haasen, eds., Elsevier Science, Amsterdam, 1996.

46.   *Journal of Phase Equilibria,* ASM International, U.S.A.

47.   *Materials Transactions A & B* see also *Trans. Met. Soc. AIME,* ASM International and TMS.

48. *Inorganic Materials*, Maik Nauka/Interperiodica.
49. *Russian Metallurgy*, Maik Nauka/Interperiodica.
50. *Russian Journal of Inorganic Chemistry*, Maik Nauka/Interperiodica.
51. *Journal of Alloy Phase Diagrams,* Incorporated into *Journal of Phase Equilibria* since 1991, Indian Institute of Metals, India.
52. *Materials Science and Technology*, Previously *Metal Science* and *Journal of Institute of Metals*, Institute of Materials, U.K.
53. *Transactions of the Indian Institute of Metals*, Indian Institute of Metals, India.
54. *Journal of the American Ceramic Society*, American Ceramic Society, U.S.A.
55. *Zeitschrift für Metallkunde*, Carl Hanser Verlag.
56. *Journal of Alloys and Compounds* Previously *Journal of the Less Common Metals*, Elsevier.
57. *Intermetallics*, Elsevier Applied Science.
58. G.C. Carter, ed., *Applications of Phase Diagrams in Metallurgy and Ceramics*, National Bureau of Standards, (U.S.) Spec. Publ. **1** and **2**, 1978, 496.
59. L. Kaufman, ed., *User Applications of Phase Diagrams*, ASM International, U.S.A., 1987.
60. F.H. Hayes, ed., *User Aspects of Phase Diagrams*, Institute of Metals, U.K., 1991.
61. User Aspects of Phase Diagrams, *Journal of Phase Equilibria,* **22**, 2001, 3 & 4.
62. N. Saunders and A.P. Miodownik, *CALPHAD - Calculation of Phase Diagrams*, Pergamon Materials Series, R.W. Cahn, ed., Elsevier Science, **1**, 1998.
63. L. Kaufman and H. Bernstein, *Computer Calculation of Phase Diagrams*, Academic Press, 1970.
64. T.G. Chart, J.F. Counsell, G.P. Jones, W. Slough and P.J. Spencer, *Int. Met. Rev.*, **20**, 1975, 57.
65. C.W. Bale and A.D. Pelton, *F*A*C*T User Guide,* 1st Edition, McGill University/Ecole Polytechnique, Montreal, 1979.
66. G. Eriksson, *Chem. Scri.*, **8**, 1975, 100.
67. B. Sundman, B. Jansson and J.-O. Andersson, *CALPHAD*, **9**, 1985, 153.
68. R.H. Davies, A.T. Dinsdale, S.M. Hodson, J.A. Gisby, N.J. Pugh, T.I. Barry and T.G. Chart, *User Aspects of Phase Diagrams*, F.H. Hayes, ed., Institute of Metals, U.K., 1991, 140.
69. B. Cheynet, *Int. Symp. On Computer Software in Chemical & Extractive Metallurgy, Proc. Metall. Soc. of CIM, Montreal'88,* **11**, 1988, 87.
70. C.W. Bale and G. Eriksson, *Canadian Metallurgical Quarterly,* **29**, 1990, 105.
71. H.L. Lukas, E.Th. Henig and B. Zimmermann, *CALPHAD,* **1**, 1977, 225.
72. U.R. Kattner, W.J. Boettinger and S.R. Coriell, *Z. Metallkde.*, **87**, 1996, 522.
73. G. Eriksson and K. Hack, *Met. Trans. B,* **21B**, 1990, 1013.
74. S.-L. Chen, S. Daniel, F. Zhang, Y.A. Chang, W.A. Oates and R. Schmid-Fetzer, *Journal of Phase Equilibria,* **22**, 2001, 373.
75. N. Saunders, X. Li, A.P.Miodownik and J-Ph. Schillé, *Proc. Symp. Materials Design Approaches and Experiences*, J.-C. Shao, T.M. Pollock and M.J. Fahrmann, eds., TMS, 2001.

REFERENCES

1. J.W. Gibbs, *The Collected Works of J. Willard Gibbs*, Longmans, Green and Co., New York, **1**, 1928, 54-371.
2. G. Masing, (Translated by B.A. Rogers), *Ternary Systems*, Reinhold Publishing Corporation, New York, 1944.
3. J.S. Marsh, *Principles of Phase Diagrams*, McGraw Hill Book Co., New York, 1944.
4. J.E. Ricci, *The Phase Rule and Heterogeneous Equilibrium*, D. van Nostrand Co. Inc., New York, 1951.
5. W. Hume-Rothery, J.W. Christian and W.B. Pearson, *Metallurgical Equilibrium Diagrams*, The Institute of Physics, London, 1952.
6. F.N. Rhines, *Phase Diagrams in Metallurgy*, McGraw Hill Book Co., New York, 1956.
7. E.M. Levin, H.F. McMurdie and F.P. Hall, *Phase Diagrams for Ceramists*, The American Ceramic Society Inc., 1956.
8. R. Vogel, *Die Heterogenen Gleichgewichte*, Akademische Verlagsgesellschaft, Geest Portig, Leipzig, 1959.
9. (a) A. Prince, *Alloy Phase Equilibria*, Elsevier, Amsterdam, London, New York, 1966. (b) A. Prince, *The Application of Topology to Phase Diagrams*, *Metallurgical Reviews*, The Institute of Metals, **8**(30), 1963, 213-276.
10. E.G. Ehlers, *The Interpretation of Geological Phase Diagrams*, W.H. Freeman and Co., San Francisco, 1972.
11. Floyd A. Hummel, *Introduction to Phase Equilibria in Ceramic Systems*, Marcel Decker Inc., New York, 1984.
12. Mats Hillert, *Phase Equilibria, Phase Diagrams and Phase Transformations, Their Thermodynamic Basis*, Cambridge University Press, Cambridge, 1998.
13. N. Saunders and A.P. Miodownik, CALPHAD (*Calculation of Phase Diagrams*), *A Comprehensive Guide*, Pergamon Materials Series, Series Editor, Robert W. Cahn, Pergamon, Elsevier Science Ltd., Oxford, 1998.
14. G. Pascoe and J. Mackowick, *Journal of Institute of Metals*, (a) **98**(253), 1970, (b) **98**(103), 1971.
15. A.D. Pelton, *Physical Metallurgy*, 3rd edition, R.W. Cahn and P.Haasen, eds., Elsevier Science, 1996, 328.
16. E.A. Alper, ed., *Phase Diagrams in Materials Science and Technology*, Academic Press, **1-3**, 1970, **4**, 1976.
17. L.H. Sperling, *Introduction to Polymer Science*, 2nd edition, John Wiley and Sons Ltd., New York, 1992.
18. A.H. Windle, *A Metallurgist's Guide to Polymers*, R.W. Cahn and P. Haasen, eds., Physical Metallurgy, Elsevier Science, **c32**, 1996, 2664-2740.
19. A.M. Donald, *Liquid Crystalline Polymers*, Cambridge Solid State Science Series, Cambridge University Press, 1992.

20. *Metals Handbook*, American Society for Metals, 1948.
21. M. Hillert, *Journal of Iron and Steel Institute*, **189**, 1958, 224.
22. H. Lipson and A.J.C. Wilson, *Journal of Iron and Steel Institute*, **142**, 1940, 107.
23. H.W.L. Phillips, *Equilibrium Diagrams of Aluminium Alloy Systems*, The Aluminium Development Association, London, 1961.
24. H. Perepezko and W.J. Boettinger, *Materials Research Society Symposium Proceedings*, **19**, 1983, 223.
25. G.M. Gulliver, *Metallic Alloys*, Griffen, London, 1922.
26. E. Scheil, *Z. Metallkunde*, 1942, 34-70.
27. W.G. Pfann, *Transactions A.I.M.E.*, **194**, 1942, 70.
28. L.E. Toth, *Transition Metal Carbides*: Refractory Materials Series, 7, Academic Press, New York, 1971 (After E. Rudy, Compendium of Phase Diagram Data, AFML-TR-65-2, part V, 1960).
29. M.B. Panish, *Journal of Physics Chem. Solids*, **27**, 1966, 291.
30. H.C. Casey Jr and M.B. Panish, *Metallurgical Transactions A.I.M.E.*, **242**, 1968, 406.
31. K.K. Shih, J.W. Allen and G.L. Pearson, *Journal of Physics Chem. Solids*, **29**, 1968, 379.
32. M.B. Panish, *Phase Diagrams*, Academic Press, New York, **3**, 1970.
33. A. Prince, *Journal of Less Common Metals*, **12**, 1967, 107.
34. A. Taylor and R.W. Floyd, *Journal of Institute of Metals*, **81**, 1952-53, 451.
35. M. Durand Charre, *The Microstructure of Superalloys*, Gordon and Breach Science Publishers, 1997.
36. J.H. Lee and J.D. Verhoeven, *Journal of Phase Equilibria*, **15**(2), 1994, 136-146.
37. K.J. Lee and P. Nash, *Journal of Phase Equilibria*, **12**, 1991, 551-562.
38. P. Willemin and M. Durand Charre, *Journal of Materials Science*, **25**, 1990, 168-174.
39. K. Zeng et al., *Intermetallics*, **7**, 1999, 1347-1359.
40. A. Taylor, *Transactions A.I.M.M.E.*, 206, 1956, 1356.
41. Chester T. Sims, Norman S. Stoloff and William C.Hagel, *Superalloys II*, John Wiley and Sons Inc., New York, 1987.
42. W.D. Forgeng and W.D. Forgeng, Jr., *Metals Handbook, Metallography, Structure and Phase Diagrams*, American Society for Metals, Metals Park, Ohio, 1973.
43. K. Bungardt, E. Kunze and E. Horn, *Archiv. f. das Eisenhüttenwesen*, **3**, 1958, 193.
44. L. Colombier and J. Hochmann, *Stainless and Heat Resisting Steels*, Edward Arnold Ltd., London, 1967.
45. Von W. Tofaute, C. Kiittner and A. Biittinghaus, *Archiv. f. das Eisenhüttenwesen*, **9**, 1935-36, 607.
46. V.G. Rivlin and G.V. Raynor, *International Metals Reviews*, **21**(1), 1980.
47. F.H. Keating, *Chromium-Nickel Austenitic Steels*, Butterworth, London, 1956.
48. Arnulf Muan, The Effect of Oxygen Pressure on Phase Relations in Oxide Systems, E.A. Alper, ed., *Phase Diagrams in Materials Science and Technology*, **2**, 1970, 1-19.

49. Arnulf Muan and E.F. Osborn, *Phase Equilibria Among Oxides in Steelmaking*, Addison-Wesley Pub. Co. Inc. Reading, Mass. and Pergamon Press Ltd. Oxford, 1965.

50. V. Raghavan and D.P. Antia, Pressure Variable, Phase Rule, and Phase Diagram: A Tutorial, *Journal of Phase Equilibria*, **19**(2), 1988, 101-105.

51. L.S. Darken and R.W. Gurry, *Journal of American Chemical Society*, **67**, 1945, 1398-1412.

52. L.S. Darken e.g., *Journal of American Chemical Society*, **68**, 1946, 798-816; **67**, 1948, 1398-1412, **70**, 1948, 2046-2053.

53. Arnulf Muan, *American Journal of Science*, **256**, 1958, 171-206.

54. Arnulf Muan, *Journal of Metals, Transactions A.I.M.E.*, **203**, 1955, 965-967.

55. V. Raghavan, *Phase Diagrams of Ternary Iron Alloys*, Systems Containing Iron and Oxygen, Indian Institute of Metals, (6), 1992, 264-277.

56. E.F. Osborn and Arnulf Muan, *Phase Equilibrium Diagrams of Oxide Systems*, The System $FeO-Fe_2O_3-SiO_2$, Plate 6: The American Ceramic Society, Edward Orton Jr., Ceramic Foundation, Columbus, Ohio., U.S.A., 1960.

57. N.L. Bowen and J.F. Schairer, *American Journal of Science*, **24**, 1932, 177-213.

58. R. Schumann Jr., R.G. Powell and E.J. Michal, *Journal of Metals, Transactions A.I.M.M.E.*, 1953, 1009-1104.

59. J.C. Yarwood, M.C. Flemings and J.F. Elliott, *Metallurgical Transactions*, **2**, 1971, 2573.

60. T.R.A. Davey, *Transactions A.I.M.M.E.*, **200**, 1954, 838.

61. H.W.L. Phillips, *Annotated Equilibrium Diagrams of Some Aluminium Alloy Systems*, The Institute of Metals, London, U.K., 1959.

62. G. Phragmen, *Journal of Institute of Metals*, **77**, 1950, 489-522.

63. COST 507, *Definition of Thermochemical and Thermophysical Properties to Provide a Database for the Development of New Light Alloys*, Office for Official Publications of the European Communities, Luxembourg, 1998, *Proceedings of the Final Workshop*, Vaals, Netherlands, **1**, 1997. *Thermochemical Database for Light Metal Alloys*, I. Ansara, A.T. Dinsdale and M.H. Rand, eds., 2, 1998, *Critical Evaluation of Ternary Systems*, Gunter Effenberg, ed., **3**.

64. Ref. 63, Al–Fe–Si, Gautam Gosh, **3**, 113-175.

65. Ref. 63, Al–Mn–Si, Alan Prince, **3**, 177-205.

66. Ref. 63, Al–Fe, **2**, 34-39.

67. Ref. 63, Al–Si, **2**, 79, 80.

68. Ref. 63, Al–Mn, **2**, 54-58.

69. H. Okamoto, *Journal of Phase Equilibria*, **14**(1), 1993, 120-121.

70. W. Fuming and H.M. Flower, *Materials Science and Technology*, **6**(11), 1990, 1082-1092.

71. H. Okamoto, *Journal of Phase Equilibria*, **14**(2), 1993, 266-267.

72. N. Saunders, *Titanium'95, Science and Technology*, Proc. 8th World Congress, P.A. Blenkinsop, W.E. Evans and H.M. Flower, eds., The Institute of Materials, 1996, 2167-2176.

73. H.M. Flower, *Materials Science and Technology*, **6**(11), 1990, 1082-1092.

74. H.M. Flower and J. Christodoulu, *ibid.*, **15**(1), 1999, 45-52.

75. T. Ahmed and H.M. Flower, *ibid.*, **10**, 1994, 272-288.

76. D.B. Miracle, M.A. Foster and G. Rhodes, *Titanium'95*, Proc. 8th World Conf., P.A. Blenkinsop, W.E. Evans and H.M. Flower, eds., The Institute of Materials, London, 1996, 372-379.

77. M. Yamaguchi, D.R. Johnson, H.N. Lee and H. Inui, *Intermetallics*, **8**, 2000, 511-517.

78. Ernest M. Levin, Carl R. Robbins and Howard F. McMurdie, *Phase Diagrams for Ceramists*, American Ceramic Society Inc., Columbus, Ohio, 1964, (Supplement 1969).

79. W.D. Kingery, H.K. Bowen and D.R. Uhlmann, *Introduction to Ceramics*, Second Edition, John Wiley and Sons, New York, London, 1976.

80. R. Kiessling and N. Lange, *Non-Metallic Inclusions in Steels*, Iron and Steel Institute, London, (2), 1966.

81. W. Schreyer and J.F. Schairer, *Journal of Petrology*, **2**(3), 1961, 324-406.

82. F. Tamas and I. Pal, *Phase Equilibria and Spatial Diagrams*, Illiffe Books, Butterworth and Co., London, 1970.

83. A. Aksay and R. Roy, *Journal of American Ceramic Society*, **45**, 1962, 229.

84. R.S. Roth, C.J. Rawn, F. Beech and J.O. Andersson, et al., *Ceramic Superconductors II*, M.F. Yan, ed., The American Ceramic Society, Westerville, Ohio, U.S.A., **2**, 1988, 13-26.

85. John D. Whitler and Robert S. Roth, *Phase Diagrams for High Tc Superconductors*, The American Ceramic Society, 1991.

86. J. MacManus-Driscoll, *Advanced Materials*, **9**(6), 1997, 457-473.

87. T. Aselage and K. Keefer, *Journal of Materials Research*, **3**, 1988, 1279.

88. J.E. Ullman, R.W. McCallum and J.D. Verhoeven, *Journal of Materials Research*, **4**, 1989, 752.

89. M. Murakami et. al. *Japan Journal of Applied Physics*, **28**(3), 1989, L399-L401.

Index

GRAPH PAPER FOR CALCULATIONS

This page can be photocopied to provide graph paper suitable for solution of the problems and for other rough calculations.

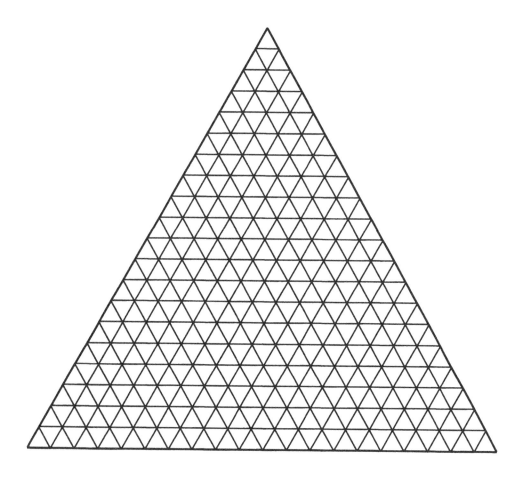

Printed and bound by CPI Group (UK) Ltd, Croydon, CR0 4YY

23/10/2024

01778245-0003